American Iron Hand Presses

AMERICAN IRON HAND PRESSES

by STEPHEN O. SAXE

Wood Engravings
by JOHN DEPOL

1992
OAK KNOLL BOOKS
NEW CASTLE, DELAWARE

Published by
OAK KNOLL BOOKS
414 Delaware Street
New Castle, DE 19720

Copyright © 1991 Stephen O. Saxe

ISBN 0-938768-35-2 (hardback)
ISBN 0-938768-36-0 (paperback)

Printed in the United States of America

*Library of Congress
Cataloging-in-Publication Data*

Saxe, Stephen O.
American iron hand presses / by Stephen O. Saxe ;
wood engravings by John DePol.
p.   c.m.
Includes bibliographical references (p.)
and index.
ISBN 0-938768-35-2 — ISBN 0-938768-36-0
1. Handpress—United States—History. 2. Printing—
United States—History. 3. Printing-press—History.
I. Title.
Z249.S23   1992
681'.62'0973—dc20   91-36587
CIP

Contents

Introduction

FROM ITS EARLIEST days in the fifteenth century until the start of the nineteenth century—a span of 350 years—the methods and equipment of printing changed remarkably little. The essential techniques of punch, matrix and hand mould, and of the wooden hand press, remained unaltered decade after decade. The only improvements of consequence to the common press between the time of Gutenberg and 1800 were the changes made by Willem Blaeu (1571-1638),[1] and these improvements were not universally adopted. The common press consisted of a heavy frame of wood; a bed to hold the form of type, which could be inked and then moved under the platen; and a bar attached to a screw for lowering the platen to make the impression. It was a simple arrangement that could be made by a joiner, and it could be repaired easily and at little cost. Although the output of the common press was limited, it was sufficient to meet the need for printed materials until the beginning of the nineteenth century, when the pace of change increased. In England the Industrial Revolution brought a large part of the population from rural farms into crowded cities; commerce expanded, and there was greater demand for newspapers, books, and all the printed matter that accompanies industry. Even in still largely rural America the population was highly literate and had a strong impulse to read and learn.

In England and America, the technology of iron working had finally arrived, and skilled artisans were building machines to do what had been done by hand. Once begun, this process of change constantly accelerated. Consider the slow pace of the world before 1800, defined by the speed of the horse and the sailing ship. A person born in that year, if he lived until 1875, would have seen the introduction of the telegraph, the steamship, the railroad, the photograph, the telephone, and the automobile.

The printing trade moved with the times. The most significant development in presses was the invention of the steam-powered cylinder press by Friedrich Koenig (1774-1833) in 1810. But parallel with this invention was the development of the iron hand press, a press which for the first part of the century was probably as important to the average person as the more complex cylinder machine. The publishing house of Harper Brothers in New York, as late as 1837, still had twenty-four hand presses in use for book printing.[2] Joel Munsell's 1850 census of printers in Albany indicated that there were twenty-nine hand presses being used in fifteen printing offices. The equipment in his own office included one Napier press, two Adams Power Presses, one Ruggles Engine Press, one card press, and four hand presses.[3] But by 1866, Thomas MacKellar wrote that "hand-presses are now restricted to country papers of small circulation, and to book-offices devoted to extra fine printing."[4]

In this book the development of the iron hand press will be presented through descriptions of more than fourteen examples, arranged chronologically. Although the title of this book refers only to American presses, I have added some significant English presses as well. The presses are all American or English iron presses with at least one example in the United States. Of these presses, only one—the Albion—was not used here for commercial printing. (The Albion seems to have come to this country only as part of the private press movement near the end of the nineteenth century.) Many of the presses were made with varying frames: there are Washington presses in acorn frames, Stansbury presses in wooden frames, and so on. What actually distinguishes one press from another is not the material or shape of the frame, interesting though these may be. It is the mechanism of the press, the means of pressing the paper against the inked type, that is important. These mechanical arrangements can be categorized as follows:

A. The screw (common press)

B. Power-increasing levers (the Stanhope and Columbian)

C. Lever-and-chill (Albion; fig. 1)

D. "Figure 4" toggle" (Washington; fig. 2)

E. Equal-length levers (Wells and Smith; fig. 3)

F. Torsion toggle (rotating inclined rods) (Stansbury; fig. 4)

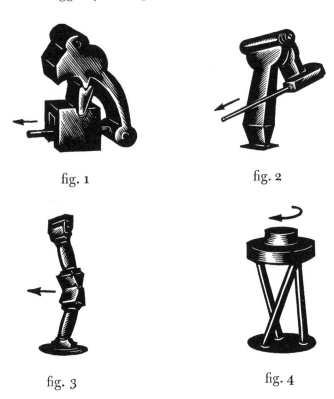

fig. 1

fig. 2

fig. 3

fig. 4

It can be readily seen from the four diagrams that although the forms may vary, the principle does not. Most rely on some form of leverage to move an inclined piece of steel into a vertical position, causing it to exert pressure (usually downward). It is this pressure that makes the impression. This method of applying pressure is more efficient than the screw, and allows somewhat faster operation and much greater force. Although these arrangements have sometimes been used inside a traditional wooden press frame, the force they create is so great that an iron frame is really required. Even iron frames have been known to crack under the strain of these powerful mechanisms.[5]

Since the development of the iron hand press is one of mechanical invention, patents are usually involved. The reader should be aware of certain basic information about early nineteenth century patents which is essential to the understanding of the subject. Patents were first issued in the United States in 1790. The system was changed slightly after 1793. From that date until 1836 the government issued patents without any examination for novelty, a policy which "resulted in many patents for things which were not new and numerous conflicting patents."[6] This explains why many of the hand presses in this book received patents when they do not seem materially different from earlier presses. Both American and British patents were granted for a period of fourteen years. By 1836 the U.S. Government was issuing patents at the rate of about 600 a year.

On Dec. 15, 1836 a disastrous fire swept the Patent Office in Washington, destroying 168 volumes of records and 9,000 patent drawings. Some of these were restored by requesting information from the inventors when possible, or by copying records at the

Holborn branch of the British Patent Office in London. But many of the most critical patents are apparently lost forever, including the earliest ones of John I. Wells and Peter Smith, which might give some answer to claims of priority between them. Most of the press patents which are missing are dated before 1822.

The Patent Office fire resulted in a reform of the laws governing patents, and those issued from then on were given numbers (prior to 1836 only a date was given). Among the other changes were examination for novelty in the claim, a seven year renewal, and, until about 1870, the requirement of a working model. Almost all of the presses in this book were patented before the 1836 reform.

For help with this book I am greatly indebted to more people than I can name in a brief space. Much basic information and photographs were supplied by the owners or curators of the presses in this book. I would like to thank for their help Ernest Lindner, Stanhope; Martin and Penny Speckter, Columbian; Donald Petty (Science Press), Ruthven; Philip Weimerskirch and Robert Oldham, Stansbury; Norman Cordes, Smith; John Jacobson (American Graphic Arts), Washington; Jeff Craemer, Union; Alexander and Ilse Nesbitt, Tufts; Walter B. Clement, Michael Phillips and William Pretzer, Foster; Eleanor Garvey, Ruggles; James Green, Bronstrup; Frank Teagle and Benjamin Mason, the press at the Shelburne Museum.

I am very fortunate to have the illustrations of John DePol to accompany the text. Indeed, they were the starting-point for the entire book. I believe they are ideally suited for this subject, being a perfect combination of artistic expressiveness and precise delineation. Moreover, the art of wood-engraving has been intimately associated with the iron hand press from the days of Thomas Bewick to the present.

I must thank especially John DePol for a joyful collaboration; Dr. James Fraser for his warm encouragement; Neil Shaver for his fine printing at his Yellow Barn Press, and Renée Weber for her constant guidance and help. Dr. Elizabeth Harris, Curator of Graphic Arts at the Smithsonian Institution, has graciously read the manuscript and made many valuable suggestions, as has Dr. Fraser. Dr. Harris and Stan Nelson of the Smithsonian have also provided information, access to their files, and countless copies of patent applications. I must also acknowledge a debt to the late Ralph Green for his pioneering *The Iron Hand Press in America*, which must be the foundation for a book on this subject. His book was the first to attempt to list and describe in detail many of these presses. I have found unpublished traces of his pioneering work in this area wherever I have turned. If I have been able to describe one or two that he was not aware of, and to provide a bit more detail on the others, I have done what I started out to do.

STEPHEN O. SAXE

NEW YORK

NOTES TO THE INTRODUCTION

1. James Moran, *Printing Presses* (Berkeley, Calif., 1973), p. 31.

2. Eugene Exman, *The Brothers Harper* (New York, 1965), p. 17.

3. Joel Munsell, *The Typographical Miscellany* (Albany, 1850; reprinted, New York, 1972), pp. 252-255.

4. Thomas MacKellar, *The American Printer* (Philadelphia, 1866), p. 212.

5. John Johnson, *Typographia*, vol. II (London, 1824), p. 536, describing the tendency of early Stanhope presses to crack.

6. *Encyclopedia Brittanica* (Chicago, 1973), p. 449.

Dedicatory Note

CONSIDERING WALTER WILKE'S monumental work on the history of the iron hand press[1] or the work of James Moran[2] and the less ambitious contributions of Herschel Logan[3] and Ralph Green,[4] one might ask is yet another work needed to celebrate this nineteenth century technical breakthrough?

When John Anderson, John DePol and I first discussed this project, using DePol's lively yet technically careful wood engravings of presses as a starting point, it seemed that, yes, there was a place for a new work. This work would not only convey the driving enthusiasm of the remarkable DePol to a new generation of hand press enthusiasts but would provide the opportunity for fresh research on the iron hand press in America. The likely author who came immediately to mind, Stephen Saxe, not only had gathered extensive notes on the iron hand press but during his editorship of the American Printing History Association's newsletter continually gave its readers the benefit of his research in this and related areas.

Three of the presses described here are held by the Florham-Madison Campus Library, a partial reason for our desire to be co-publisher with Neil Shaver's Yellow Barn Press. (Shaver had been involved in a number of printing and publishing projects with DePol throughout the 1980s. Their compatibility insured effective design and production of the projected book.) Another reason was our involvement with these and other individuals here in Madison in various book arts activities, and joint publication would allow public recognition of those individuals who have given our library presses in the past or otherwise encouraged the use of the iron hand press. To Elaine Rushmore Brown (1916-1988), Virginia Haberly, Emma Linen Dana, Mary Christiansen, Delight Rushmore Lewis and Ginna Johnson Scarry this book is dedicated.

JAMES FRASER

FAIRLEIGH DICKINSON UNIVERSITY

MADISON, NEW JERSEY

NOTES

1. *Die Entwicklung der eisernen Buchdruckerpresse.* (Pinneberg, Ger.: Verlag Renate Raecke, 1983)

2. *Printing Presses.* (Berkeley and Los Angeles: University of California Press, 1973)

3. *The American Handpress: its origin, development and use.* (Whittier, CA: Zoller Press, 1980)

4. *The Iron Hand Press in America.* (Rowayton, CT: self-published, 1948)

The Stanhope Press

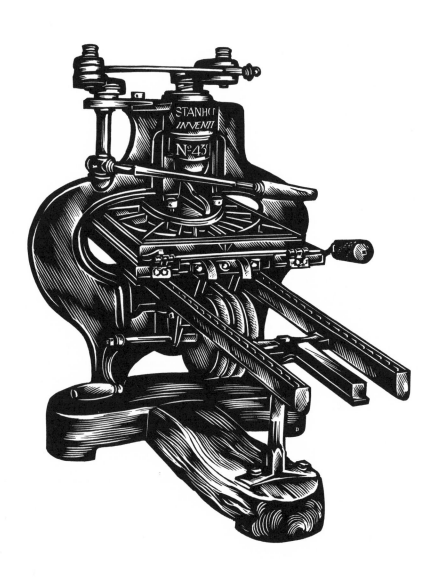

The Stanhope Press
1800–ca.1840

THE ERA of the iron press began at the start of the nineteenth century in England, home of the Industrial Revolution. Although there had been precursors—presses using metal parts as substitutes for wood—it was the Stanhope press that dramatically changed the hand press and pointed the way to the future.

It was, after all, the age of the machine—the steam engine, the power loom, the spinning jenny. England had coal to smelt iron and the skilled mechanics to work the iron. Charles, the third Earl Stanhope (1753-1816) devoted his life to science and technology, and especially to attempts to improve techniques of printing. Of his many projects, only his iron press was a lasting contribution.

With the help of machinist Robert Walker, Lord Stanhope built a press made of one massive iron casting about 1800. The shape of the frame resembled the iron frame of a press made by Wilhelm Haas of Basel in 1772. In the Stanhope press, the familiar screw of the common press was kept, but the power of the press was multiplied many times by the use of compound levers. These power-increasing levers and an entirely iron construction were Stanhope's great contribution. The press can thus be seen as a transitional one, retaining elements of the past but using materials and methods of the future. Because he desired to benefit the printing industry, Lord Stanhope did not patent his press and thus made it available to all.[1] The first press made by him was used by the great printer Bulmer; it later was in the office of the J.B. Nichols printing company until the premises were leased to His Majesty's Stationery Office in 1939. The press has not been seen since, although a photograph of it survives.[2]

The earliest users of the press included the *Times* of London, which used a "battalion" of Stanhope presses[3] during the early years of the nineteenth century, and William Bulmer, mentioned above. The *Times* appreciated the speed of the presses, and Bulmer their control and strength of impression. That strength of impression was considerable. On the common press two pulls of the bar, with a shifting of the bed between pulls, was necessary to print a form. The physical effort required was such that pressmen developed a characteristic enlargement of the right shoulder muscles that was an occupational badge.[4] The rate of printing on a Stanhope press was from 200 to a maximum of 250 impressions an hour,[5] but full forms could now be printed at a single pull. It is not surprising to learn that early Stanhope presses were often likely to crack at the middle of their cast iron frames.[6] Lord Stanhope then produced a more massive design, the Stanhope press "of the second construction." This press, which appeared about 1806, had widely rounded cheeks, in contrast to the straight sides of the first press. Hansard, describing the press, wrote "that when an experienced pressman first tries it, he cannot feel any of the reaction which he has been accustomed to, and will not believe, till he sees the sheet, that he has produced any impression at all . . . "[7]

The increase in power that levers "on Stanhopean principles" gave to the screw convinced some printers that they could apply these levers to the wooden press. Abraham Rees wrote, in *The Cyclopaedia*, in 1819: "Several Stanhope presses, that is presses having levers to work the screw, have been made in wood by Mr.

Brooke, or altered from old wooden presses, but though this is an improvement upon them, it is greatly inferior to the iron frame."[8] A press of this kind that has survived is on view at St. Bride's Institute, London. Most of these presses eventually failed; the enormous strain could not be withstood by wooden members.[9]

The Stanhope press was such a clear advance that it soon attained great popularity. It was exported to most European countries as well as to America; in 1811 David and George Bruce of New York (later to become important typefounders) had one in use.[10] The press also began to be manufactured in many countries, including France, Germany, and Italy. It is even possible that Francis Shields, a pressmaker who emigrated from London to New York, may have manufactured Stanhope presses in America about 1810-11.[11]

The Stanhope press attained a certain celebrity as a symbol of the power of the press. It is described by Honoré de Balzac in his *Illusions Perdues* (1837-1839), and it is pictured by George Cruikshank in his illustrations for the political satires of William Hone attacking the forces of censorship and reaction. Honoré Daumier used it in his satirical lithographs for the same purpose in France.

Although the earliest Stanhope presses were made by Robert Walker, Lord Stanhope's policy of not patenting the invention eventually ensured at least five other makers.[12] Three presses "of the first construction," i.e., with straight cheeks, have survived, as well as a great many Stanhopes of the later design. There appears to be only one full sized one in the United States, however, and that is the press illustrated here, "of the second construction." It is part of the Ernest A. Lindner collection of antique printing machinery at the International Museum of Graphic Communication, Buena Park,

California. It was acquired by Mr. Lindner in London. Another Stanhope press is at McGill University in Montreal. A charming table-top version of the Stanhope, a French press called "La Typote," can be seen in the Buena Park collection.

The Lindner press stands 5′–2″ tall, bears the serial number 439 and the hand-chiseled inscription *Stanhope Invenit* and *Walker Fecit*. It is estimated that the press was made about 1810. Within a decade after that date the Stanhope press began to fall from favor as the competing Columbians (and still later, Albions) appeared. The Stanhope press was manufactured until about 1840, but was still advertised in Harrild's London printing machinery catalogues as late as 1860.

NOTES FOR CHAPTER ONE

1. Johnson, *Typographia*, vol. II (London, 1824), p. 536.

2. Horace Hart, *Charles Earl Stanhope and the Oxford University Press* (Oxford, 1896; reprinted London, 1966 with notes by James Mosley), p. xxiii.

3. Moran, *Printing Presses*, p. 54.

4. Milton R. Hamilton, *The Country Printer* (New York, 1936), p. 45.

5. Colin Clair, *A History of Printing in Britain* (New York, 1966), p. 211; *Abridgments of Specifications Relating to Printing* (London, 1859), p. 23.

6. Johnson, *Typographia*, vol. II, p. 536; Moran, *Printing Presses*, p. 50.

7. T.C. Hansard, *Typographia* (London, 1825), p. 425

8. Abraham Rees, *The Cyclopaedia*, Vol. 28, "Printing," (London, 1819).

9. Hansard, *Typographia*, p. 647.

10. Rollo G. Silver, *The American Printer 1787-1825* (Charlottesville, Va., 1967), p. 48.

11. *Long-Island Star*, Oct. 23, 1811; Moran, *Printing Presses*, p. 52; Silver, *American Printer*, p. 48.

12. Hart, *Charles Earl Stanhope* (2nd ed.), p. xxvi.

The Columbian Press

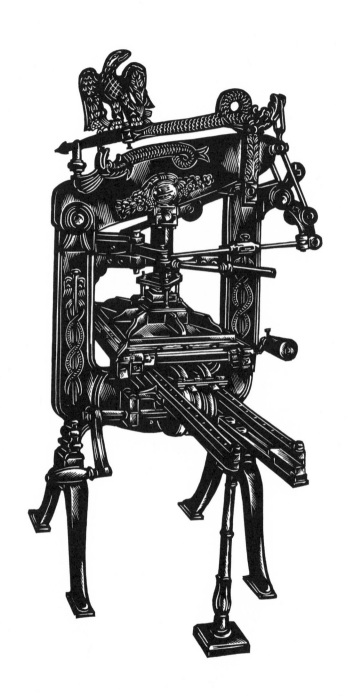

The Columbian Press
1813−ca. 1913

THE COLUMBIAN PRESS represents the first flowering of Yankee talent and genius applied to the printing press. Its inventor, George Clymer (1754-1834), was born in Philadelphia. He became in turn a teacher, carpenter, cabinetmaker and inventor-mechanic, best known by 1801 for his improved pump used in the construction of the first permanent bridge across the Schuylkill. During the same period he also began experimenting with improvements to the common press, but his earliest ideas were not radical departures from tradition. In 1805 he was first listed in Robinson's *Philadelphia Directory* as a printing press manufacturer.[1]

Although at least two Stanhope presses had been imported to the United States from England (one was in the printing office of David and George Bruce in New York, 1811)[2] there is no evidence that Clymer ever saw them. According to the first American edition of *The Cyclopaedia*, published in Philadelphia, "In the year 1800 Mr. Clymer began to make improvements on the construction of the common printing press. In the course of his investigations of the different modes of applying power, he successively employed principles similar to those used by Mr. Medhurst [rotating inclined rods] and Mr. Roworth [inclined planes.] The first was entirely abandoned at an early stage of his labors; but his presses constructed on the principles of inclined planes have, for several years past, been in general use, and are esteemed by printers as very superior in power and equality of impression to the common press."[3] Clymer eventually came to use a series of compound levers. His arrangement of power-multiplying levers was capable of exerting enormous pressure with the lightest effort, as will be testified to by anyone who has ever pulled the bar of a Columbian. The patriotic cast-iron eagle at the top of the press acted as an adjustable counterweight to raise the platen after an impression. (Other early iron presses, such as the Stanhope and the Wells, also used counterweights, which were later superseded by springs.)

Clymer's press was introduced to the public in 1813. Its great advantage was that it could print easily a very large form at a single, light pull. According to the American edition of Rees' *The Cyclopaedia*, "so slight a degree of strength is necessary to print a royal sheet on this press, that it completely removed from the printing business the long standing reproach of press-work being destructive of health and life."[4] A skillful pressman for the New York *Commercial Advertiser*, working a Columbian with a partner, could print 250 sheets an hour on one side, in what was considered a fine performance.[5] We know that nearly all the newspapers in New York used the Columbian[6]—newspapers had to meet the needs of reaching a mass audience quickly then as now—but its acceptance in the developing United States was limited. The Columbian's great cost, $300 to $500,[7] at a time when a new Ramage press could be had for $130,[8] was one obstacle. Its great weight and the difficulty of making repairs were others. Even Clymer's flair for publicity—reflected in the ornate and fanciful patriotic motifs of the castings—could not overcome the resistance of American printers. We may estimate that no more than 25 of the presses were made in America, used by printers in New York and

Philadelphia, and one each in Hartford and Albany.[9] Not one of these American-made Columbians is known to have survived.

In 1817 Thomas Barnitt of Philadelphia either bought Clymer's business or became his agent. His name, listed as "Columbian printing press manuf. and turner in brass, iron, etc." appears in the city directories of 1819 through 1824.[10] Columbians were clearly obtainable for a few years after Clymer's departure for England.

In May, 1817, at the age of 63, George Clymer set sail for London. There, he was granted patent no. 4174 on November 1, 1817, and he arranged for an English press-maker, R.W. Cope, to manufacture the press.[11] The earliest English-made Columbian (and hence the earliest Columbian) that survives is dated 1819. It is located at Simpson Printers, Swindon, Wiltshire, England.[12] The press found immediate favor among English printers, in spite of its American flavor, for it worked easier and had a larger printing surface than the Stanhope. By 1841 Savage could write, "It ranks in the opinion of practical men, generally, as the next in estimation to the Stanhope press."[13] One reason for this ranking was that although more expensive, the Stanhope was slightly faster.

Clymer publicized his press with testimonials from leading printers. After a test in St. Petersburg in 1818 the Columbian was awarded a prize of 600 rubles by Czar Alexander I and soon after, a gold medal worth 100 ducats by the King of the Netherlands.[14] Favorable notices also came from the writers of printers' manuals; Johnson, in his *Typographia*, reported that a Columbian press on which he had made some improvements was able to print a large wood engraving that could not be printed on a Stanhope for lack of power—the power of the impression "greatly exceeded our most sanguine expectation," and "astonished every beholder—even Mr. Clymer himself."[15] Hansard, in his *Typographia*, took a somewhat sneering attitude toward the press's embellishments, but had to say that it performed well.[16]

By 1830 Clymer entered into partnership with Samuel Dixon, trading as Clymer and Dixon. When Clymer died in 1834 at the age of 80, Dixon continued making the presses, but they continued to bear the name of Clymer and Dixon. After 1845 the name was changed to Clymer Dixon & Co.[17] Before the middle of the century the press was well established in England; in the years after the patent expired (ca. 1838) at least twenty other manufacturers began making it. Variations were made as well in Germany, France, the Netherlands and Belgium. But the press was hardly known in America, where the Washington press held sway.

The Columbian was made in ten sizes, ranging from Foolscap Folio (15″ x 9–3/4″) to Extra Size Double Royal (42″ x 27″). The earliest Columbian presses made by Clymer had the bar on the off-side, away from the pressman. Hansard stated that the bar was moved to the near side about 1825,[18] and this is confirmed by the fourth-oldest existing Columbian press, dated 1825, owned by Paul B. Quyle, Jr. of Murphys, California.[19]

Eventually the Columbian was overtaken in popularity by the Albion, made by R.W. Cope (Clymer's original press maker) after 1825. The Albion, which used a spring to raise the platen, was slightly faster in operation though it required more exertion; in the printing trade, then as now, speed of presswork is often all-important.

Two of the many makers of the Columbian, Harrild & Sons and Frederick Ullmer, both of London, made them well into the first decade of the twentieth century.[20] By this time they were mainly sold as proof presses; as late as the 1970s used Columbians were still being advertised in *The British Printer* under the heading "proof presses."

There are quite a few Columbians in England, Canada and the United States. They may be found at the Smithsonian Institution, the International Museum of Graphic Communication (Buena Park), the University of Nevada (Reno), the University of Texas (Austin), South Street Seaport Museum (New York), California State University (Northridge) and Massey College (Ontario, Canada) among other places. The press illustrated here is owned by Penny Speckter of New York. Martin and Penny Speckter, through the good offices of Beatrice Warde, purchased the press in 1968 from its original owner, Stanbrook Abbey of Worcester, England.[21] It was purchased for the Abbey in 1876 at a cost of £25, and became the first press of the noted Stanbrook Abbey Press, operated by the nuns of the Benedictine order. The press was made by (or for) Reed & Fox, typefounders of London.[22] It is Crown size (the platen measures 23″ x 19″) and stands 6′–0″ high. It is believed to be the oldest press in continuous use as a private press.

NOTES FOR CHAPTER TWO

1. James Robinson, *The Philadelphia Directory for 1805* (Philadelphia, 1805).
2. Silver, *American Printer*, p. 48.
3. Abraham Rees, *The Cyclopaedia* (Philadelphia, 1810-1824), vol. 29.
4. *Ibid.*
5. *One Hundred Years* (Philadelphia, 1896), p. 28.
6. Munsell, *The Typographical Miscellany* (Albany, 1850), p. 126.
7. Silver, *American Printer*, pp. 48-49.
8. Milton R. Hamilton, *Adam Ramage and His Presses* (Portland, 1942), p. 3.
9. James Moran, "The Columbian Press," *Journal of the Printing Historical Society* No. 5 (London, 1969), p. 5.
10. Jacob Kainen, *George Clymer and the Columbian Press* (New York, 1950), pp. 21-22.

11. E.C. Bigmore and C.W.H. Wyman, *A Bibliography of Printing* (London, 1884), vol. i, p. 343.

12. Moran, "Columbian Press," p. 20.

13. William Savage, *A Dictionary of the Art of Printing* (London, 1841), p. 172.

14. Munsell, *Typographical Miscellany*, pp. 138, 140.

15. Johnson, *Typographia*, vol. II, pp. 548-9.

16. Hansard, *Typographia*, p. 656.

17. Moran, "Columbian Press," p. 9.

18. Hansard, *Typographia*, p. 656.

19. Moran, "Columbian Press," p. 20.

20. *Ibid.*, p. 11.

21. Letter from Martin Speckter, 5 October 1986.

22. Benedictines of Stanbrook, *The Stanbrook Abbey Press*, (Worcester, 1970), pp. 17, 160, [163].

The Ruthven Press

The Ruthven Press
1813–ca.1830

IN THE EARLY YEARS of the Republic there were strong ties between Edinburgh, Scotland and Philadelphia among the members of the printing trade. At that time Philadelphia was, with Boston and New York, one of the three great centers of commerce in the United States. Among the Scots who emigrated to Philadelphia about 1795 were the typefounders Archibald Binny and James Ronaldson, David and George Bruce (soon to move on to New York) and the successful press manufacturer, Adam Ramage.

After making a great success of his own Ramage press—an improved common press well suited for both town and country printing offices—Ramage heard of the new press invented in Edinburgh by the Scots printer John Ruthven. Ruthven's press, patented in Britain in 1813, had several novel features. The bed of the press, holding the type form, was fixed and stationary. The platen was mounted on springs above two wheels on either side. After inking the form, the platen was rolled into place over the form. Instead of pulling the bar sideways in the usual manner, the pressman pushed down on the bar at the front of the press, allowing the weight of his body to help in making the impression. Clamps on the side of the press pulled the platen downward, against the action of the springs, until the impression was made. Then the springs raised the platen, which was rolled away from the form, and the tympan and frisket raised to remove the printed sheet and to receive the next.

The arrangement was novel and highly logical, though it was not always easy for pressmen to adjust to the new movements called for by the Ruthven press. The pressing down of the bar set in motion a series of compound levers under the press and made the impression. The levers moved the platen rapidly at first, but it slowed appreciably as the impression was made and the greatest power was brought to bear. The iron mechanism of the press was mounted within a square wooden arrangement of legs and side members, all very compact and occupying minimum space. With no frame above the platen, the press was much lower than the usual hand press, its highest point no more than about three and a half feet from the floor.

In 1818 Adam Ramage began manufacturing and selling the press in Philadelphia, probably under a license agreement with Ruthven. The press was described by Van Winkle of New York in 1818 in his *Printers' Guide*, though he had not yet seen it.[1] Ruthven's fellow Scot, Abraham Rees, writing in his *The Cyclopaedia* in 1819, stated that "the levers beneath the table are well contrived to have the best effect in saving time, and producing an immense pressure . . . the machinery is well contrived in all its parts, both for performance and stability . . . it is far cheaper than the Stanhope press."[2] We do not know how many Ruthven presses Ramage sold in the United States. The virtues of the press included its ability to produce a good, even, powerful impression. But it was never wholly accepted by many printers, and several of the English writers of printers' manuals found fault with it. Johnson, writing in his *Typographia*,[3] objected to it because he claimed that the pressman's hand might slip from the bar while pulling an impression, allowing the bar to fly upward and injure his wrist or arm. (Ralph Green has written that if the bar should fly up "it was quite likely to knock his teeth out, an accident which was not at all uncommon."[4] We do not know his authority for this statement.) Johnson also claimed that it was nearly impossible to oil or clean the levers under the bed without disassembling the entire press.[5] Savage called it "a good and powerful press," but added that "the head and platen are heavy and require exertion to push them back off the form."[6] This was made more difficult because the tracks along which the platen rolled were inclined to make it easier to push over the bed, but harder to push it away.

Green's chronology of iron press manufactures in the United States has the Ruthven no longer being made after 1821,[7] only three years from its introduction. We have a statement by Thomas F. Adams of Philadelphia, in his printers' manual of 1837, "It was introduced into this country about sixteen years ago, by Mr. Adam Ramage . . . but owing to the introduction of rollers shortly after, they were abandoned, not conveniently admitting the application of rollers to them; they were, however, much esteemed for doing fine work."[8] An automatic inking roller apparatus had been developed for use with the standard iron hand press and was made by Hoe, among others, for use where high productivity was required. It is known that Ruthven patented his own inking roller device, but it does not seem to have made a difference.

There is only one known surviving Ruthven press in the United States. It was mentioned by Hamilton in 1942 as being "in the office of Rhode Printing Company, Kutztown, Pennsylvania"[9] and by Green in 1948 as "still being used for proofing in a small eastern Pennsylvania printing office."[10] That office was the same Rhode Printing Company. Rhode employees believed that the press was purchased from a newspaper in Orwigsburg, Pennsylvania. Rhode was bought by Hughes Printing Company and renamed Craftsmen, Inc. In 1950, two years after the purchase, workers making structural changes in the basement of the building uncovered what appeared to be "a pile of iron junk."[11] Eventually the press was restored and put on display at the offices of the Printing Company of America. PCA was bought by American Can Company and the press was displayed in its steel-and-glass corporate headquarters in New York. In 1976 it returned to eastern Pennsylvania, to Science Press in Ephrata. It is now part of the Russell C. Hughes Museum at Science Press, and it is this press that is illustrated here. The platen is 24″ x 29–1/2″ and the bed is about 3–0″ from the floor.

NOTES FOR CHAPTER THREE

1. Cornelius Van Winkle, *The Printers' Guide* (New York, 1818), pp. 196-201.
2. Abraham Rees, *The Cyclopaedia* (London, 1819), vol. 28, "Printing."
3. Johnson, *Typographia* (London, 1825), vol. II, p. 545.
4. Ralph Green, *The Iron Hand Press in America* (Rowayton, Conn., 1948), p. 24.
5. Johnson, *Typographia*, vol. II, p. 545.
6. Savage, *Dictionary of the Art of Printing*, p. 719.
7. Green, *Iron Hand Press*, p. 28.
8. Thomas F. Adams, *Typographia* (Philadelphia, 1837), pp. 325-326.
9. Hamilton, *Adam Ramage*, p. 38.
10. Green, *Iron Hand Press*, p. 24.
11. Bob Tipton, "The Ruthven Press," *Science Profile* (Ephrata, Pa., 1985), p. [12].

The Wells Press

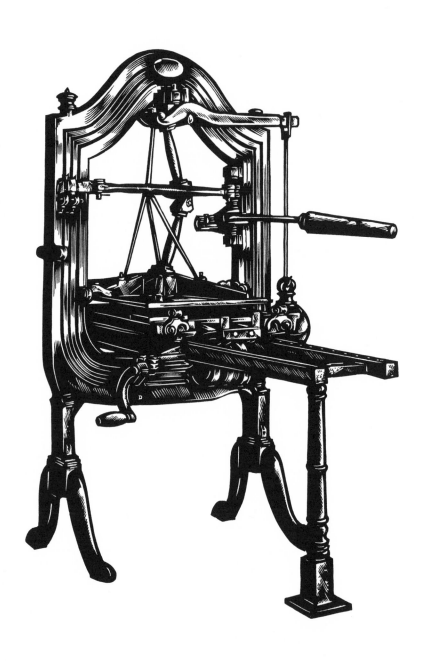

϶ 4 ϵ

The Wells Press
1819–1833

JOHN I. WELLS (1769-1832) was a true Connecticut Yankee, a mechanic who saw a better way to make something work and who turned his idea into reality. His idea was to replace the screw and the compound lever with the toggle as a means of transmitting power for the printing press. In that, he was a true pioneer.

Wells was born in East Hartford, Connecticut. He became a cabinetmaker, working as a journeyman in a New York shop before returning to Hartford to start his own business. His shop prospered and he began to branch out into making mattresses. By 1807 he was supplying ink to printers and had a patent for a mill to grind paint pigment. Both his inks and paints were of very high quality, and gradually became more important than his furniture shop. He devised and patented a press with a lever toggle joint to extract linseed oil for paint, and his association with printers through his ink business suggested to him that this toggle could substitute for the screw of the common and Ramage presses.[1] He obtained a patent on an improved lever printing press February 8, 1819. Unfortunately, this key patent is now lost, but its contents have been described (see page 25). This was the first American press not to use the screw or compound levers (as the Columbian, Stanhope and Ruthven presses did).

In an advertisement in 1820, Wells stated that he had first introduced his "Lever Printing Press" in 1815. He claimed that three journeymen at the Great American Bible Office in New York had such high regard for his press that, rather than use the presses on hand, had spent their own money ($50 a year) to lease three of his presses for their work. Wells offered his presses in three sizes: the Medium, $300; the Super Royal, $325; and the Imperial $350.[2]

In a statement published in a printers' manual in 1828, Wells wrote "in the *History of the Printing Press* it will probably be admitted as a fact, that my experiments proved the *first successful effort* to introduce the power of Levers, *end-wise*, into the printing press." He added, "My experiments commenced in 1816, and for 2 or 3 years wooden frames strengthened with iron, were used; but proving too elastic, iron frames were substituted."[3]

In 1822 Wells admitted that structural weaknesses in the frame of his early press had appeared, but he added that in all cases he had redesigned the part and replaced it free of charge. He also stated that the presses made by both Abraham Stansbury and Peter Smith infringed on his patent. He said that Stansbury had once lived in Hartford near his shop and frequently watched the manufacture of the press. In the case of Smith, he said that many of the parts of that press were identical with his own, and there were statements by several printers that Smith had carefully examined the Wells press before producing his own. Wells personally called on both men to warn them of the infringement,[4] apparently without result. Yet there does seem to be some merit in his position. Both his competitors' presses were on the scene for many years, and the Smith press, made by R. Hoe & Co., long after the Wells press.

In 1825 Wells announced that through various improvements and greater manufacturing efficiency, he was able to reduce his prices to $220, $230, and $240 for the three sizes he made. He now had from four to six workmen assembling them, as well as agents in

Boston, Philadelphia, Albany and Utica. He noted that the presses were boxed and could be shipped if necessary.[5] We can readily conclude that he was a successful manufacturer, and this is borne out by records of his real estate purchases in Hartford.

A good description of the Wells press appeared in the *Connecticut Mirror* of August 2, 1819:

> . . . The frame, platten, and several other parts are of cast iron; and the weight of the cast and wrought iron is about 1500 lbs. The power is obtained by two upright levers, footing in the centre of the platten; within a strong circle upon the plate. These levers are fifteen inches in length, one and three fourths of an inch square in body, and four inches wide at the ends. They move in sockets of the semi-circle of half an inch; falling back in the centre, two inches, from a perpendicular line—this admits of the rising of the platten. They are governed in this joint, and forced nearly to a straight line, by two horizontal levers, attached in connection with the arm or bar, to the back line of the press . . . The platten is raised by a spindle, suspended upon a balance lever, by a balance weight . . . [6]

Early Wells presses used a heavy iron ball as a counterweight. This awkward method of raising the platen after an impression was changed to springs; other makers' presses with springs had by now appeared. Another improvement he effected in his early presses was to move the bar from the off-side of the press to the near side, after printers complained of the extra reach needed.

The Wells press illustrated here has the heavy counterweight ball and the bar still located on the off-side. It is in the collection of the Smithsonian Institution and is one of the earliest Wells presses surviving. (Slightly earlier is press No. 50, dated 1823, at the Worcester

Historical Museum, Worcester, Massachusetts.) [7] The Smithsonian press is 6'-0" high, and has a platen measuring 20-3/4"x30-1/2". It was given to the Smithsonian Institution by the American Type-founders Company in 1915.[8] The brass plate affixed to the head of the press reads:

John I. Wells

Patent Lever Press No. 54

Hartford Conn.

Another Wells press of later date is on view at the printing office of Mystic Seaport, Mystic, Connecticut. It has springs to raise the platen instead of the older counterweight.

The Connecticut Historical Society has a brass plate taken from a Wells press with no. 246 and the name of Wells' son, J. Hubbard Wells.[9] The press to which the plate was once attached was probably made after Wells' death on April 12, 1832. Soon afterward the Wells press ceased to be made, having been surpassed by other inventions; yet his introduction of the toggle lever was a major advance.

NOTES FOR CHAPTER FOUR

1. Houghton Bulkeley, "John I. Wells, Cabinetmaker-Inventor," *Connecticut Historical Society Bulletin*, vol. 26, no. 3 (Hartford, July, 1961), p. 68.

2. *Ibid.*, p. 70.

3. [Thomas F. Adams], *An Abridgment of Johnson's "Typographia"* (Boston, 1828), pp. 302-303.

4. Bulkeley, "John I. Wells," pp. 70-71, quoting Middlesex *Gazette* (Middletown, Conn., 20 June 1822).

5. *Ibid.*, p. 71.

6. Quoted by Silver, *American Printer*, pp. 50-51.

7. Letter from Elizabeth Harris, Smithsonian Institution, 6 October 1987.

8. Communication from Stan Nelson, Smithsonian Institution, 20 November, 1987.

9. Bulkeley, "John I. Wells," p. 71.

The Stansbury Press

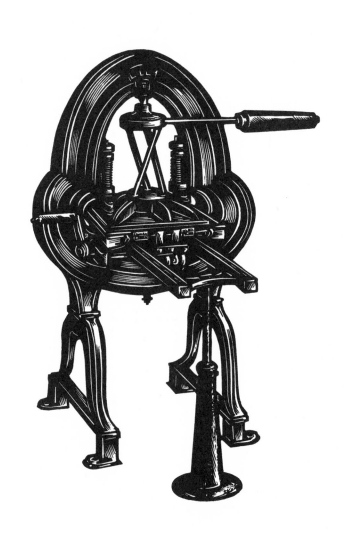

ꝫ 5 ꝫ

The Stansbury Press
1820–1885

THE CAREER of the Rev. Abraham Ogier Stansbury (1776-1829) was as diverse as it seems possible to imagine. He was, at various times, a bookseller and publisher, grocer, one of the first superintendents of a school for the deaf in the United States, the first person to make a lithograph in New York, ordained minister, and inventor.

He was born July 19, 1776 in Philadelphia but most of his life was lived in and near New York. He first appears in Longworth's New York directory as a merchant in 1801.[1] In that year and in the two following he became a publisher and bookseller, sometimes in partnership with one of his brothers and sometimes alone. He was the publisher of an edition of Nathaniel Bowditch's *The New American Practical Navigator*.

In 1804 or 1805 he went to England, and there received a patent for a lock he had invented. After his return to America in 1807 he became an ordained minister and in 1817 began a short career as superintendent of schools for the deaf in Hartford and New York that lasted until about 1820.[2] Some of his sermons and talks made during that period exist; one was "On the lawfulness of lotteries, and the propriety of Christians holding tickets."[3] In 1821, no doubt drawing on his observations of printing during his career as a publisher, he obtained a patent for a printing press employing a toggle mechanism different from the mechanisms of the Columbian, Ruthven, and Wells presses which had preceded it.

The mechanism, known as the "torsion toggle," consisted of two straight rods attached at the lower end to a circular plate. At the upper end the rods were fixed to the head. When the bar was pulled the circular plate rotated, and the lower parts of the inclined rods with it, thus bringing the rods into a vertical position and forcing the platen downward to make the impression.

Stansbury received his American patent on April 7, 1821, but it is doubtful whether his torsion toggle was original. At some time before the patent was issued, probably about 1816, George Medhurst of Soho, London, a clockmaker and ironfounder, devised a toggle on the same principle. Medhurst's toggle was installed in a common press, and was later described by Hansard as having "considerable merit."[4] We have no way of knowing whether Stansbury knew of the Medhurst invention, which was not patented, but it is quite possible. Medhurst's toggle used short rods, while Stansbury's were considerably longer.

In researching the Stansbury press, Robert W. Oldham has found five surviving presses,[5] but there are more, if we take into consideration the later Stansbury presses made by Isaac Adams and R. Hoe and Co. after 1845 and 1867 respectively. The earliest of the Stansbury presses seems to be the one at Richmondtown Restoration, Staten Island, New York. A parallel with the Medhurst press is seen at once, since this press is essentially a wooden common press with the Stansbury toggle in place of the screw. This press differs from all the other Stansbury presses in most details, including its use of a cast iron ball for a counterweight, similar to the early Wells press at the Smithsonian. It is possible that this press was made by Stansbury in New York, after which the rights to manufacture it were given to the Cincinnati Type Foundry. As early as 1822 that company advertised "STANSBURY'S PATENT PRESSES and SCREW PRESSES made to order; a specimen of each may be seen at the office of the Liberty Hall."[6]

The earliest known Stansbury made by the Cincinnati Type Foundry is in the collection of the Missouri Historical Society in St. Louis. It, too, is essentially like a common press in form, and, though the toggle mechanism is missing, it clearly was a Stansbury. The Cincinnati Type Foundry then began producing a hybrid form with a wooden common press frame and a decorated cast iron head. Eventually some of the common press features began to disappear, giving these Stansbury presses a distinctive look. This form of press can be found at the Henry Ford Museum in Dearborn, the Ohio Historical Society in Columbus, and the Methodist Publishing House in Nashville.

According to Oldham, the Cincinnati Type Foundry offered the Stansbury press in an all-wood as well as a hybrid wood-and-iron frame, but between 1829 and 1834 converted the press to an all cast iron frame. After the Stansbury patent expired in 1845, a version of the press was made by Isaac Adams, an important press inventor and manufacturer in Boston. One of these presses, in an acorn frame and with three rods rotating at the upper end (like the later Hoe version), can be seen at the Updike printing collection at the Providence Public Library.

In 1867 the firm of R. Hoe & Co. in New York began manufacturing a Stansbury press very similar to the Adams style. Hoe was always an aggressive company, and it had a history of manufacturing whatever products it thought would sell. The Hoe Stansbury press was sold from 1867 to 1885 (and perhaps even as late as 1894; in that year *The American Dictionary of Printing and Bookmaking* noted that "no other hand press is made, with the exception of the

Stansbury, used for hat tips . . . "[8] It had an all-iron frame in the acorn shape. The press made by Hoe differed from the earlier Cincinnati presses in other respects. Like the Adams version, it employed three inclined rods, instead of two; it dispensed with the spindle; and the bar was moved to the upper end of the rods, which were attached to a rotating disc. The press was only made in one size.

Hoe Stansbury presses are owned by the Dunstan Press, West Scarborough, Maine, and Phil Cade, Winchester, Massachusetts. The one pictured here is at Fairleigh Dickinson University, Madison, New Jersey. This press is of special interest because it was used by Loyd Haberly, at one time director of the Gregynog Press, for his private press printing in the United States. Haberly has written, "My last press—now in Fairleigh Dickinson University's Friendship Library (Madison Campus)—was bought at a Boston sale of the effects of Thomas Bird Mosher, the Portland, Maine, printer . . . I brought this press to Rutherford where I printed one book in the University's original Castle building, and two other small ones in the University house that has been my home for twenty years."[9] It came to the Friendship Library in 1973 when Professor Haberly retired from his teaching and other duties, and donated his papers, a collection of his printing, and his press to the Library. It has a very small platen measuring only 13″ x 16″ (the only size made by Hoe) and is 5′–0″ tall. The forestay support shown is undoubtedly not original.

NOTES FOR CHAPTER FIVE

1. *Longworth's American Almanac, New-York Register, and City Directory* (New York, 1801), p. 283.

2. I am indebted to Philip Weimerskirch of the Updike Collection, Providence Public Library, Providence, Rhode Island for unpublished information about the

career of Abraham Stansbury. Much of this material was researched by Dr. Weimerskirch and presented by him in a talk given at the Old Southeast Church, Brewster, New York, on 11 September 1983. Rev. Abraham O. Stansbury was the eighth pastor of the church.

3. Abraham O. Stansbury, *Considerations on the lawfulness of lotteries, and the propriety of Christians holding tickets* . . . (New York, 1813).

4. Hansard, *Typographia*, p. 652.

5. Robert Oldham, "Abraham O. Stansbury and the Torsion Toggle Hand Printing Press," *The Chronicle of the Early American Industries Association*, Vol. 36, no. 4 (Albany, December, 1983), pp. 72-76. Oldham's article is the best source of information about the Stansbury press.

6. Walter Sutton, *The Western Book Trade* (Columbus, 1961), p. 14, quoting the Cincinnati *Inquisitor*, Feb. 19, 1822.

7. Oldham, "Abraham Stansbury," pp. 75-76.

8. *The American Dictionary of Printing and Bookmaking* (New York, 1894), p. 256.

9. Loyd Haberly: *Loyd Haberly, Poet & Printer* (Madison, New Jersey, 1972), p. 11. This booklet is a keepsake consisting of remarks by Loyd Haberly on the occasion of the presentation of his Stansbury press to the Library of Fairleigh Dickinson University, printed by John Anderson at The Pickering Press. It was reprinted in *The Printing Art*, vol. 1, no. 3 (London, Autumn, 1973), pp. 2-13.

The Smith Press

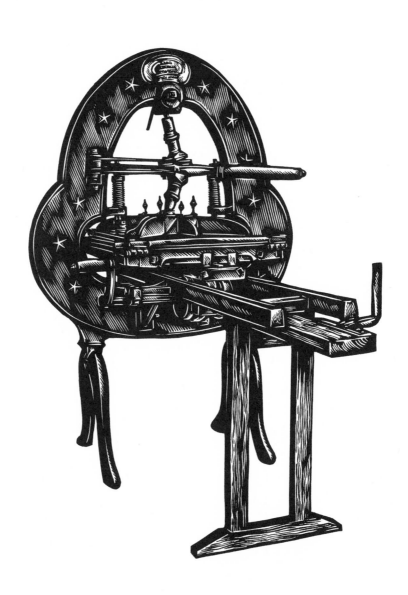

The Smith Press
1821–1890

PETER SMITH (1795-1823) was one of the few American press inventors to receive a college education (Yale, 1816). He was the younger brother of Matthew Smith, a carpenter and printers' joiner who in 1805 formed a partnership with Robert Hoe. Peter was independent of Smith, Hoe & Co., operating a series of cabinetmaking shops until about 1820, when he succeeded to the partnership on the death of his brother.[1] Smith, Hoe & Co. made "wood goods" for printers, that is, type cases, frames, and other wooden equipment including common presses.

Peter Smith's first press patent is dated December 29, 1821, but it has been lost. All we know of it is a mention by his competitor, John I. Wells, who wrote " . . . His Patent is dated '29th Dec. 1821,' of which I have a copy. He claims that his 'improvement consists chiefly in applying the *Wedge Power*,' whereas all his power is derived from his Levers."[2] Wells had patented his toggle joint in February of 1819, (this patent is also lost) and he stated that Smith had closely observed his presses in New York in 1820. Wells accused Smith of patent infringement.[3] In that judgment it is easy to agree, since Smith's toggle joint is very similar to Wells'. Smith obtained another patent, dated April 6, 1822, which is available. It shows a new kind of pressure-applying mechanism. The bar of the press pulled a wedge of steel between two pivoting quadrants; when the thick end of the wedge was between them, the platen was forced downward. Although this press was probably never manufactured, it is remarkably similar to the press patented by David Barclay in England on July 26, 1821. Barclay's press is illustrated in James Moran's *Printing Presses*.[4] Barclay stated that the idea for his press was communicated to him "by a certain foreigner residing abroad." In this press the power is applied by a wedge being drawn between two rollers.[5] Moran surmises this foreigner to be George Clymer of Philadelphia, but we believe it more likely that it was Peter Smith. This possibility is strengthened by the fact that the Barclay press is shown in a near-acorn frame, and Peter Smith did introduce that form in the United States and his press has always been associated with it.

Although Wells' patent might well have been infringed, the patent laws of the day did not require originality, and the Smith press had the advantage of being made and sold by the aggressive firm of Smith, Hoe & Co. in New York, a large commercial center, while Wells remained in Hartford. Smith's press had shorter toggle members than Wells', and was mounted in a massive cast iron acorn frame, attractively decorated with relief castings of stars. Although Green has written that the press "contained no new features worthy of comment,"[6] the frame was certainly new and distinctive. The press appeared on the market about 1821. It was reliable and strong, and was highly successful.

Peter Smith did not enjoy the success of his new press for long. He probably retired from the partnership in 1822 and died the following year. The firm of Smith, Hoe & Co. became R. Hoe & Co., the most successful press manufacturer in America for the rest of the nineteenth century and a good portion of the twentieth. The Smith press, with modifications, was in their catalogues until 1890.

After the introduction of Rust's Washington press, the Hoe company recognized that the acorn frame was structurally inferior. Hoe then enclosed the Smith toggle in the rectangular Washington frame, but the distinctive and attractive acorn frame remained popular (and, in fact, is still much sought after today).

A number of Smith acorn presses have survived; examples can be found at the Art Students' League in New York, the University of California in Santa Cruz, Smith College in Northampton, and Old Sturbridge Village in Massachusetts. The press illustrated here is owned by Norman Cordes, Cordes Printing, Glen Rock, New Jersey. It originally came from Perry County, Pennsylvania, where it is believed to have printed the first newspaper in that area. From there it went to a New Bloomfield, Pennsylvania printing office, where it was used to print posters and where Mr. Cordes found it in 1972. The press is 5′–5″ tall and has a platen measuring 21″ x 30″. It bears a brass plate which reads:

Peter Smith's Patent

No. 410

Manufactured by

Rob't. Hoe & Co.

New-York

Although it is not dated, it is estimated that the press was made about 1835.

NOTES FOR CHAPTER SIX

1. Frank E. Comparato, *Chronicles of Genius and Folly* (Culver City, 1979), p. 33.
2. [Thomas F. Adams], *An Abridgment of Johnson's "Typographia"* (Boston, 1828), pp. 302-4.
3. Bulkeley, "John I. Wells," pp. 70-71, quoting Middlesex *Gazette* (Middletown, Conn., 20 June 1822).
4. Moran, *Printing Presses*, p. 75.
5. *Abridgments of Specifications Relating to Printing* (London, 1859), p. 156.
6. Green, *Iron Hand Press*, p. 13.

The Washington Press

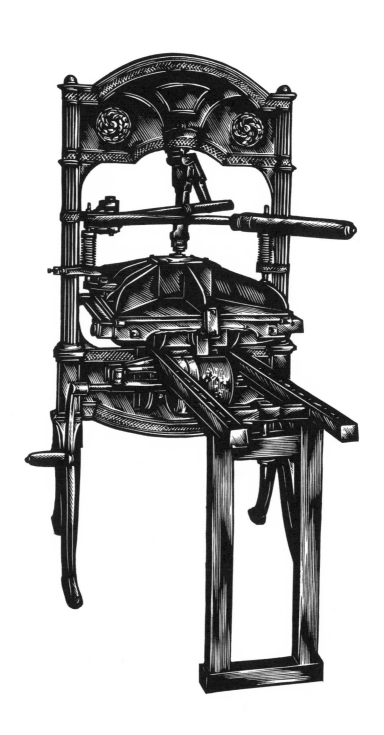

The Washington Press
1821–*ca.* 1910

SAMUEL RUST, the inventor of the Washington press, is almost unknown today, although his press was by far the most successful hand press in America, and one that is still widely used. Rust was a New York printer who seems to have moved from one location to another almost every year between 1817 and 1820. In 1819-20 he was listed in Longworth's New York directory as "printer and grocer,"[1] which seems to indicate that he had to augment his income. On May 13, 1821 he patented his first press. This patent has been lost, as were so many from this period, but as Ralph Green suggests, "we may infer that this patent covered his 'figure 4 toggle . . . ' "[2] About 1821 Rust formed the partnership of Rust and Turney (probably James Turney, a printer) to make his presses. Brass plates on early presses attest to the partnership. The earliest Rust presses contained the distinctive toggle in a massive cast iron frame. An early press in the collection of the Henry Ford Museum in Dearborn has the toggle in this heavy oval frame, decorated with stars and urns. An engraved brass oval on the press indicates that it was made by Rust and Turney,[3] and it seems to be one of the earliest Rust Washington presses surviving. Another of the same style is at the Oregon Historical Society.[4] Early wood engravings of 1828 show the press with the traditional acorn frame, looking very much like a Smith press except for the mechanism.[5]

Rust's second patent of April 17, 1829 is available. The Rust &

Turney partnership was dissolved at about this time. The drawing accompanying the patent shows a Washington press virtually as it is today. One of Rust's seven claims in the patent, and one that turned out to be very important, was his new form of frame. It consisted of two horizontal cast iron beams for the top and bottom of the press, held in place by cylindrical side posts. Although the patent does not describe the construction in detail, in practice these side posts were hollow and contained rods that could be bolted to the top and bottom beams. This frame was lighter and stronger than the massive acorn frame, and had the additional advantage of being able to be disassembled for shipment. In a large country with uncertain transportation, this was a major factor in the success of the Washington. Another was the "figure 4" toggle, so named from its three pieces in that configuration. It was more powerful than the Wells toggle, and less likely to jump out of its socket.

The great success of Rust's press soon brought him to the attention of R. Hoe & Co., of New York. Hoe, as maker of the Smith press, was Rust's chief competitor, and it soon became apparent that the Rust press was superior structurally and mechanically. Several times Hoe tried to buy the rights to Rust's patent, but without success. In 1835 Matthew Smith of the Hoe company wrote to another member of the firm:

> . . . About a week ago we sent John Colby [a Hoe employee] to Rust to learn on what terms his patent could be bought. Knowing that R. would be disposed to take all the advantages of *us* he could if he thought *we* wanted to purchase so we plan'd it that Colby should go to him on his own a/c & pretend that he was going to leave us & start opposition: this pleased R. very much & he was quite anxious to sell . . . [6]

Colby was able to buy Rust's entire establishment: patent rights, shop lease, patterns, stock, tools, platen lathe & other lathes, horsemill and horse, and Rust's obligation not to compete for a term of eight years, all for $3,000. According to Smith, at that time Rust was making about forty presses a year. Shortly afterward Colby moved everything to the Hoe plant[7] and from 1835 on the Washington press was one of the Hoe company's main products. Over the years Hoe & Co. made several thousand Washington presses. By 1870 serial number 5,400 had been reached, although there were gaps in the numbering sequence each year. The press was made until well after the turn of the century. Rust retired from the printing field to become a manufacturer of patent lamps.

But Hoe & Co. was by no means the only manufacturer of the Washington press. When the Washington patents expired many others rushed to produce their own versions. Adam Ramage's American press was one; the Cincinnati Type Foundry, A.B. Taylor Mfg. Co. (New York), Franklin Type Foundry (Cincinnati), Palmer & Rey type foundry (San Francisco), and the Marder, Luse type foundry (Chicago) all produced Washington presses. Washington presses were seen in printing offices all over the country; they were the most popular iron hand press, by far.

Long after the hand press was obsolete for commercial printing, the Washington press had a use. When photo-engraving was developed in the 1890s the Washington press was found to be ideal for proofing blocks. Beginning about 1895, massive presses were made expressly for proofing photo-engravings by Paul Shniedewend & Co. of Chicago (The Reliance), F. Wesel Mfg. Co. of New York (the Wesel), Morgans & Wilcox of Middletown, New York and the Ostrander-Seymour Co. of Chicago. Many more of these later presses are in use today than are the earlier presses. Although they lack the grace of the earlier presses and have no history of production printing, they are functionally the same.

The two illustrations are representative of the many varieties of Washington presses. The engraving on page 42 shows a Rust press of early (but not the earliest) construction. Although it has no markings, it was probably made by Rust or by Hoe soon after the latter took over Rust's establishment. It is now in the antique press collection of American Graphic Arts, Elizabeth, New Jersey. The platen is 23″ x 36″ and the press is 6′–0″ high.

The frontispiece engraving shows the Washington press as made by R. Hoe & Co. in mid-century. It is decorated with bas-relief medallions of Benjamin Franklin and George Washington in the circles occupied by rosettes in the earlier press. It is the typical Washington press. The platen measures 37″ x 24″ and it stands 6′–1″ tall from the floor to the top of the brass finials. This press belonged to Arthur Rushmore, proprietor of the noted Golden Hind Press. Mr. Rushmore, for twenty-seven years a book designer for Harper Brothers, used the press to design and try out page layouts. Many of the Golden Hind Press books and ephemera were printed on this press. Mr. Rushmore's family gave the press to Fairleigh Dickinson University, where it is now located in the Friendship Library on the Madison campus. The press has cast on the head :

WASHINGTON PRESS
R. HOE & CO.
NEW-YORK
No. 4159

The serial number, 4159, indicates that the press was made by Hoe in 1860.[8]

Ralph Green has written, "The name Washington, as applied to the iron press, has been in common use for over a century, but Samuel Rust has been forgotten. He invented and perfected a tool which, if its long and useful life is considered, has been of more importance and greater help to the country printer than any one item in the office."[9]

NOTES FOR CHAPTER SEVEN

1. *Longworth's American Almanac, New-York Register and City Directory* (New York, 1820), p. 342.

2. Green, *Iron Hand Press*, p. 16.

3. Communication from William S. Pretzer, Henry Ford Museum, 22 July 1987.

4. A photograph of this press (mislabeled a Ramage) is reproduced in Robert F. Karolevitz, *Newspapering in the Old West* (Seattle, 1965), p. 132.

5. [Thomas F. Adams], *An Abridgment of Johnson's "Typographia"* (Boston, 1828), p. 306.

6. Comparato, *Chronicles*, pp. 34-5.

7. *Ibid.*, p. 35.

8. From a list of Hoe-made Washington press serial numbers supplied by R. Hoe & Co. and published in *The American Pressman* (Chicago, November, 1965), p. 49A.

9. Green, *Iron Hand Press*, p. 22.

The Albion Press

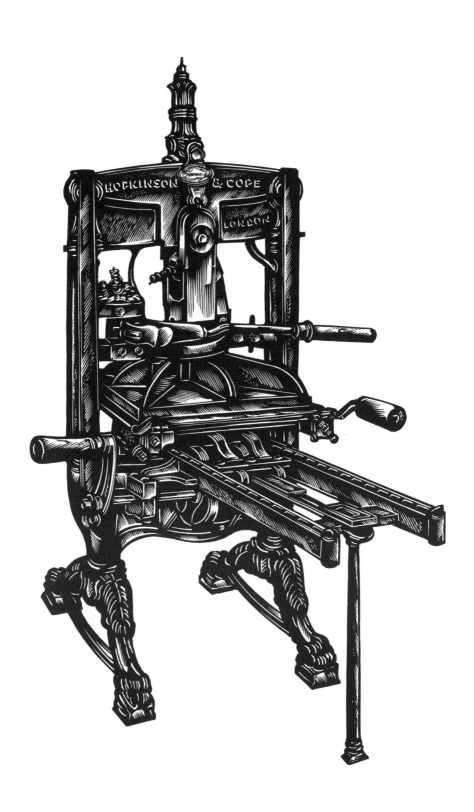

⟫ 8 ⟨

The Albion Press

1822–*ca.*1985

RICHARD WHITTAKER COPE (?-1828) is a shadowy figure, and
what we know of him is gleaned from trade directory listings and
similar mundane references. He probably started in the business of
making and selling printing equipment as early as 1815. It was
claimed by Bigmore and Wyman that Cope was employed by
Clymer to manufacture Columbian presses for his first ten years in
England.[1] This statement has not been verified. Evidence exists to
indicate that Cope's Albion press was made as early as 1822,[2]
and even its name suggests a competition with the Columbian
that had since 1817 established a firm foothold in England.

Cope was the first to use a form of toggle joint in England, and he
probably did so independently of John Wells of Connecticut,
whose patent is dated 1819. Cope's press had a very compact
mechanism, enclosed inside a hollow piston. When the bar was
pulled, it forced a lug, (later, a chill) into an upright position,
thus driving the platen down. This arrangement was much more
compact than the long levers of American presses like the Wells. A
single spring housed in the brass cap above the head of the press
served to raise the platen after an impression, and gave the Albion a
handsome and distinctive look.

The Albion was far different from the Columbian in its decora-
tion. Instead of the flamboyant ornamentation of the latter, the
Albion had classical motifs much in vogue at the time: Doric pillars

and pediments, rosettes, lozenges. The press was far simpler than the Columbian. John Johnson, writing in his *Typographia*, was enthusiastic about the simplicity, ease of working, and power of the Albion. He wrote, "There are so few parts belonging to it, and consequently the machinery is in itself so simple that there is not the least chance of their being put out of order, or liable to the least accident from wear."[3] Johnson's illustrated description of the press remains one of the most complete.[4]

In 1827 Cope introduced a form of the press that had counterweights in the form of classical urns or the royal arms, which moved along a bar at the top of the press, similar to the eagle on a Columbian. Of the hundreds of Albions in existence today, only a few with the counterweights are known and it is safe to say this type of press was abandoned quickly. Around 1828 J. Sherwin and J. Cope introduced the Imperial, a press that was very similar to the Albion but which had some improvements in the mechanism. These changes were soon incorporated in an improved Albion of 1830. The relationship between J. Cope and R. W. Cope, if any, is unknown. There are Imperial presses at Columbia University's School of Library Service and at Massey College, Ontario.

By the time Cope died in 1828 the Albion was quite popular in Britain, and often preferred to the Columbian. One major advantage was that it worked faster than the Columbian, although slightly more effort was required because the Albion had fewer levers through which the power was transmitted. But speed in production has always been important in the printing trade, and eventually the Albion surpassed all other hand presses in England.

After Cope's death the business was continued by John Hopkinson, who improved parts of the press and made them less susceptible to breakage. The firm was styled "Hopkinson & Cope."

Hopkinson died in 1864, although the firm's name remained for many years.[5]

During the 1850s and 1860s many manufacturers rushed to produce this popular press. More than a dozen companies made them in Britain, and there were others in France, Czechoslovakia, Belgium, and Australia. Some of the best-known British firms were Harrild, Ullmer, Miller & Richard, Caslon, Stephenson Blake, Notting, Sherwood, and D. & J. Greig. Dawson, Payne & Lockett in Wharfedale, Yorkshire, made them as late as the 1940s. Even more recently a 12" x 7–1/2" table model was produced by Ullmer of London in the 1980s.

The Albion press has been associated with the private press movement in England since its beginning. Albions were used by one of the earliest, the Daniel Press in Oxford, as well as by the great triumvirate of private presses: William Morris' Kelmscott Press, Walker and Cobden-Sanderson's Doves Press, and St. John Hornby's Ashendene Press. The press illustrated here is one of these, with a distinguished provenance. It was manufactured by Hopkinson & Cope in 1891 and given the serial number 6551. Morris bought it in 1894 for £52.10s and it became one of the three Albions in use at the Kelmscott press.[6] It was reinforced with iron bands around the frame to withstand the formidable pressure needed to print the large, black forms of the Kelmscott *Chaucer*. Morris' secretary, Sir Sydney Cockerell, has written, "We had three Albion presses at the Kelmscott Press, and they were just like any other Albion presses without any magical attributes."[7]

After Morris' death the press was owned successively by C. R. Ashbee's Essex House Press, the Old Bourne Press, and the Pear Tree Press, before it was bought by Frederic and Bertha Goudy in 1924. The Goudys brought it to America where it became part of their Village Press. Later the press was at Spencer Kellogg's Aries Press and then Melbert Cary's Press of the Woolly Whale in New York.[8] In 1961 the press was acquired by Dr. J. Ben Lieberman, the well-known exponent of letterpress printing in America, and it remains in the home of his widow, Elizabeth Lieberman.

The press is massive. It is 5′–8″ tall, with a platen 29″ x 21″. In spite of Sydney Cockerell's admonition, there is something magical about this press. It has been part of too many important private presses to be just another press.

NOTES FOR CHAPTER EIGHT

1. E.C. Bigmore and C.W.H. Wyman, *A Bibliography of Printing* (London, 1884), vol. 1, p. 343.

2. Reynolds Stone, "The Albion Press," *Journal of the Printing Historical Society*, No. 2 (London, 1966), p. 63. This article is the best source of information about the Albion.

3. Johnson, *Typographia*, Vol. II, pp. 553-557.

4. *Idem.*

5. Reynolds Stone, "The Albion Press," pp. 70, 71.

6. "The Kelmscott/Goudy Press" (Oakland, Calif, 1987) p. 3. (Keepsake.)

7. Reynolds Stone, "The Albion Press," p. 58.

8. "The Kelmscott/Goudy Press," p. 3.

The Union Press

The Union Press
ca. 1826–?

ONE OF THE most obscure presses to emerge from the machine shops of Boston in the second quarter of the nineteenth century was the Union press. Facts about the press are scarce and references to it often seem to conflict. Perhaps the most solid information we have can be found on the brass plate affixed to the only surviving Union press, which reads:

UNION PRESS
No. 40
Manufactured by
E. BARTHOLOMEW
for
Greele & Willis
BOSTON
1826

We can be grateful that the plate has not been removed by a souvenir-hunter, as many have been, for it gives us a good deal of information that is not found elsewhere.

We can assume, for example, that the press was not patented, since that would certainly have been mentioned if it were. We can further assume that with a serial number as low as 40, manufacture must have begun not long before 1826.

Samuel Greele and Henry Willis, for whom the press was made, by 1828 were the owners of the New England Type Foundry, which began operations in Boston in 1824. In 1826, the date on the press, the Boston directory gave the owners of the foundry as [John] Baker & Greele.[1] It is not until 1828 that Greele and Willis are listed as partners in the typefoundry.[2] We can only assume that there was a working relationship for at least two years before the listing. Type foundries were usually suppliers of presses and other equipment to the trade, and in this instance we may surmise that Greele & Willis were supplying the presses to Baker & Greele's typefoundry, which in turn sold them to printers.

The key name on the plate is clearly E. Bartholomew. Erastus Bartholomew (1783-1860) was born in Goshen, Connecticut and became an inventor and machinist. He worked with the celebrated William Church (1778-1863), press and composing machine inventor, in Vershire, Vermont and in Boston. Among their joint productions was "the first breech loading gun," "the first machine for the making or forging of nails," and one of the earliest knitting machines in the country. On his own he obtained patents for making cordage and chain cables.[3] The Bartholomew family gene-alogy states that Isaac Adams was an apprentice of his, and that the Adams Press was "in a measure the joint invention of the two."[4] The press referred to was probably the highly esteemed Adams Power Press, a bed-and-platen machine, which was manufactured by Bartholomew.

Yet we find ourselves on slippery ground here. *Ballou's Pictorial Drawing-Room Companion*, which printed a biography of Adams in 1855, states flatly, "In 1826 he invented and built the hand printing-press called the 'Union Press.' This was the first of his printing-presses. Large numbers of them were manufactured and sold by Erastus Bartholomew."[5] Since Bartholomew was alive at the time of the biographical sketch and could have refuted any misstatement, it was probably correct.

Reconciling the apparently conflicting information, we seem to have a picture of master machinist Bartholomew and apprentice Adams in a close and harmonious relationship, with Adams beginning to invent successful new presses in 1826 and Bartholomew manufacturing them. Adams had a long and extremely successful career and acquired a considerable fortune. In later years he manufactured, under his own name, an iron hand press very similar to the Smith press. Bartholomew was listed as a "machine builder" in the Boston directories during the period 1825-1830.[6]

The Union press, although much like earlier presses, has some points of interest. The mechanism is the usual equal-length lever toggle invented by John Wells, but on the Union press the bar has a geared connection with the rod leading to the toggle. This might indicate a connection with the 1827 patent of Samuel Couillard, Jr., which had a friction wheel rotating against a cam on the bar of the press. Couillard was part of the small group of press inventors and makers in Boston at the time. Another point of interest is the shape of the frame, which has straight sides narrowing to a point at the top, in the shape of a Gothic arch. The Union press frame is decorated with thirteen stars, symbolic of the American Union. In most other respects the press presents no original or unusual features.

From the serial number, forty, we cannot jump to the conclusion that at least forty Union presses were made, since many firms had gaps in their numbering system. In any case, only this one is known to have survived. It was bought by Edward Smith at an auction of the assets of the Tuttle Printing Company of Plattsburgh, New York.[7] Smith stored it in a Quonset hut behind a Howard

Johnson Motor Lodge in Plattsburgh. The press had probably been the property of Lawrence Larrabee, operator of the Tuttle Printing Company, and it has been speculated that Larrabee obtained the press from an old Plattsburgh firm, the Sentinel Publishing Company.

Smith sold the press to printing equipment dealer and collector Alan Dietch, who in turn sold it to Jeff Craemer of San Rafael, California. It was on exhibition at the Smithsonian Institution from 1980 to 1983, when it was returned to Mr. Craemer's private printing museum in San Rafael. The press has a platen that measures 20–1/4" x 27–1/4", and is 5'–9" tall.

NOTES FOR CHAPTER NINE

1. *The Boston Directory* (Boston, 1826) , p. 36.
2. *The Boston Directory* (Boston, 1828) , p. 126.
3. George Wells Bartholomew, *Record of the Bartholomew Family* (Austin, 1885) , p. 174.
4. *Idem.*
5. "Isaac Adams, Inventor of the Adams Power Press," *Ballou's Pictorial Drawing-Room Companion* (Boston, 24 February, 1855) , p. 124.
6. *The Boston Directory* (Boston, 1825, 1826, 1827, 1829) .
7. Letter from Elizabeth Harris, Smithsonian Institution, 6 October 1987.

The Tufts Press

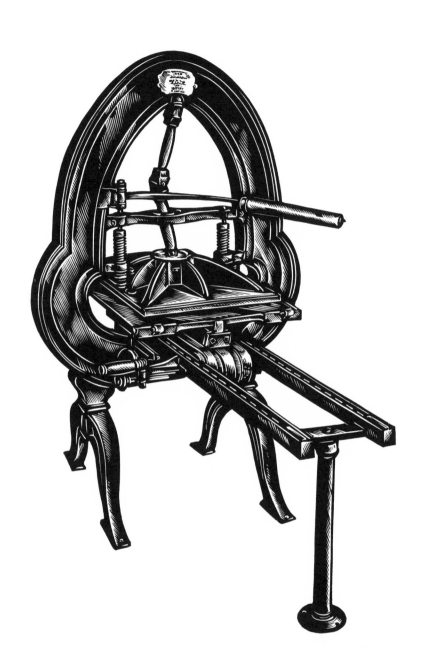

The Tufts Press
1831–1837

ONLY A FEW Tufts presses have survived the years. Dr. Elizabeth Harris has records of only seven.[1] There may be a few more undiscovered in the New England area.

Otis Tufts was a Boston machinist. He worked for Phineas Dow[2] and later for another Dow apprentice, Isaac Adams (1802-1883),[3] inventor of the Adams Power Press and other presses. There is indeed a remarkable concentration of press-inventing and press-making activity to be seen in Boston in the 1820s, including Daniel Treadwell (1791-1872, press inventor, silversmith, and Harvard professor), Erastus Bartholomew, Samuel Couillard, Jr., Phineas Dow, Isaac Adams, and Otis Tufts. In 1829 Tufts began to be listed in the city directory as a machinist in his own right.[4]

In the 1830s machinists were often press-makers. Tufts began to make his version of Treadwell's power press in 1834; he later built a six horsepower steam engine to drive the Adams Power Presses in the shop of the Boston printer, Henry O. Houghton.[5]

Tufts was granted two patents, on July 30 and November 7, 1831. The specifications for both seem to be lost; only the drawings can be found in the Patent Office archives. Both sets of drawings relate to a compound lever action at the end of the bar to effect the straightening of the toggle joint by a pushing action. Ringwalt's *American Encyclopaedia of Printing* mistakenly lists a Tufts patent for 1813, but that is a typographical error for 1831.[6]

Tufts' press was a close copy of Peter Smith's Hoe press, complete with acorn frame. Its toggle was strikingly similar to Smith's short equal-length levers, except that the knuckle joint was on the left of the press instead of on the right. A pull of the bar pushed a connecting rod against this knuckle, forcing it into a vertical line. The Smith press, as well as Isaac Adams' version, pulled the knuckle joint straight. From a distance, all three presses, with their acorn frames, could well be mistaken for each other. The irony of these copies of Smith's press is that his toggle mechanism itself was probably a copy of John Wells' equal-length lever toggle.

Another connection with Isaac Adams of Boston was the fact that the Tufts press, at least for the latter part of its history, was manufactured by Adams. The Tufts press remained popular in New England as late as 1894.[7]

The press illustrated here is located at the Third and Elm Press, conducted by Alexander and Ilse Buchert Nesbitt in Newport, Rhode Island. Alexander Nesbitt is a distinguished designer, typographer, and calligrapher, and is the author of *The History and Technique of Lettering*.[8] The Nesbitts relate that the press once belonged to John Howard Benson, a noted American letterer and stonecutter who published two books on lettering and calligraphy.[9] After his death, the press was taken to a junkyard. Mr. Nesbitt heard of the press's fate, and with Benson's son searched for and found it in 1965. It is surmised that Benson originally obtained the press from a local newspaper office.

The press has its original brass marking plate, which reads

Patent

No. 23

Invented and Manufactured

by

OTIS TUFTS

Boston

The serial number, 23, does not relate to the patent. It does indicate, however, that this was one of the earliest of these presses made, probably about 1831-2. Of course, we have no way of knowing how many more were produced; the chances are, not many. The press has a platen that measures 22″ x 30″. The Nesbitts use the press actively for their letterpress printing, which means that this press has had an active working span of over 160 years with no sign of faltering.

NOTES FOR CHAPTER TEN

1. Letter from Elizabeth Harris, 6 Oct. 1987.

2. Comparato, *Chronicles*, p. 19.

3. *Idem.*

4. *The Boston Directory* (Boston, 1829) , p. 267.

5. Comparato, *Chronicles*, p. 21.

6. J. Luther Ringwalt, *American Encyclopaedia of Printing*, (Philadelphia,1871), p. 236.

7. *The American Dictionary of Printing and Bookmaking* (New York, 1894) , p. 256.

8. (New York, 1950) .

9. *The Elements of Lettering* (with A.G. Carey, Newport, 1940) and *The First Writing Book*, a calligraphic translation of Ludovico Arrighi's *Operina* (New Haven, 1954) .

The Philadelphia Press
—
The Bronstrup Press

⟩ 11 ⟨

The Philadelphia Press
1834–1850
The Bronstrup Press
1850–1875

ADAM RAMAGE (1772-1850) emigrated to the United States from
Scotland in 1795, at the age of twenty-three. With him on the trip
was George Bruce, then fourteen, later a leading New York type-
founder. Within a few years of Ramage's arrival in Philadelphia
he had established himself as a "printers' joiner," a builder and
repairer of wooden presses.[1] After 1800 he began to improve these
presses by enlarging the diameter and reducing the pitch of the
screw, thus increasing the power of the impression.[2] He became
widely known as the maker of inexpensive, durable, well-made
wooden presses that were ideally suited for smaller country newspaper offices. After 1820, as the iron press came into use, he improved his wooden press with iron parts.[3] But iron presses, although heavier and more expensive, were far stronger and easier
to operate. It was not until 1833 or 1834 that Ramage introduced
his own iron press, the Philadelphia press.

Although his press's toggle (as illustrated in Adams' *Typographia*, 1837) appears to be like the Wells/Smith toggle, it also has
some of the characteristics of Rust's "figure 4" Washington toggle.
Ralph Green has called it "a compromise between the Wells type
and the Washington, in fact it was similar to Rust's 'figure 4'

toggle, stretched out to appear to be like the simple Wells arrangement."[4] The patent for the press, dated November 19, 1834, makes no claim for the mechanism, but does claim originality in the form and materials of the frame. Instead of the massive, brittle cast iron of other presses, Ramage used wrought iron in the shape of a triangle at the top of the press. By utilizing the principle of the truss instead of the solid beam, Ramage succeeded in doing for the iron press what he had already done for the common press: he made it simpler, lighter, stronger, and cheaper. Ramage claimed that "the whole press is not half the weight of the cast iron presses, and is so constructed that a man can carry each of the pieces, with the exception of the bed. The price will be less than the cast iron presses, and they will be warranted."[5]

In view of Ramage's claim that the pieces, except for the bed, could be carried by one man, I believe that Green's statement that "the frame could not be dismantled for transportation"[6] is incorrect. A photograph of the Philadelphia press used to print the *Deseret News* (Salt Lake City) in 1850 seems to show a frame attached to the rest of the mechanism by bolts.[7] The triangular frame was made from wrought iron bar stock, about 4″ wide and 1″ thick.

Although it seems to have merited success in terms of its strength, lightness, and low cost, the Philadelphia press did not gain the great popularity of Ramage's wooden presses. But there was great competition in the marketplace by 1833, and printers had several highly satisfactory, established presses to choose from. In assessing how many Philadelphia presses were sold, we run into a problem of nomenclature. Because both his wooden presses and the Philadelphia press were popularly known as "The Ramage Press," it is hard to distinguish the two in written accounts. "Ramage" presses were used to print the first newspapers in several Western states, but often we cannot be sure which press was meant. We can only say that a fair number of Philadelphia presses must have been made, since they were produced over a span of more than forty years. By the year 1837 Ramage was reported to have manufactured over 1,250 presses of all kinds.[8]

When Adam Ramage died in 1850, his successor, Frederick Bronstrup (1811-1900), continued his line of presses. This line included the Philadelphia press, which now (slightly modified) was renamed the Bronstrup press. Bronstrup continued to manufacture it until about 1875. It was offered in three platen sizes: 16″ x 22″, 20″ x 26″, and 22–3/4″ x 29–1/2″.[9] Bronstrup, a native of Germany, was naturalized in 1840, and was originally listed in city directories as a blacksmith, and later as a printing press maker and machinist.[10]

Adam Ramage is chiefly remembered for his improved wooden presses, which were by far the most popular in the early years of the nineteenth century. But Lawrence Wroth has written that "Ramage's wooden press improvements were only a small part of his achievement. His numerous inventions for the improvement of the iron press established his fame as one of the great press builders of the first half of the nineteenth century."[11]

One of the iron presses he manufactured was the American press, invented by Sheldon Graves. It is illustrated and briefly described by Adams as "a combination of the toggles of the Washington Press and the lever or elbow of the Smith."[12] From this description and the illustration accompanying it we can see a close relationship with the toggle of the Philadelphia press. None of the American presses seem to have survived.

Very few of the Philadelphia or Bronstrup presses have survived, and we may take this as a reflection of the press's only modest success. A Philadelphia press is owned by the *Deseret News*, Salt Lake City, where it was used in 1850 to print the first issue of that newspaper.[13] Bronstrup presses are owned by the Museum of Science and Industry, Chicago; Bowne and Co., Stationers, New York; and the Smithsonian Institution. The Bronstrup press illustrated is the Bowne example, now on long-term loan to Fairleigh Dickinson University, Madison, New Jersey. Bowne acquired it in the late 1970s from The Stonehand, at that time a dealer in printing antiques in New York.

This press has the largest platen, 22–3/4″ x 29–1/2″, and measures 6′–4″ high to the top of the center finial. The brass plate contains no serial number, but reads

F. Bronstrup

Successor to A. Ramage

Philadelphia.

NOTES FOR CHAPTER ELEVEN

1. Hamilton, *Adam Ramage*, pp. 2-3.

2. *Ibid.*, pp. 10-13.

3. Thomas F. Adams, *Typographia*, 3rd edition (Philadelphia, 1845), p. 272.

4. Green, *Iron Hand Press*, pp. 24-25.

5. Thomas F. Adams, *Typographia*, 1st edition (Philadelphia, 1837), pp. 328-333.

6. Green, *Iron Hand Press*, p. 25.

7. Karolevitz, *Newspapering*, p. 148.

8. Adams, *Typographia*, 1st ed. (1837), p. 328.

9. Ringwalt, *American Encyclopaedia of Printing*, p. 81.

10. Information about Frederick Bronstrup in a letter from James Green, The Library Company of Philadelphia, 22 December 1987.

11. Lawrence C. Wroth, *The Colonial Printer*, 2nd ed. rev., and enl. (Portland, Me., 1938) p. 86.

12. Adams, *Typographia*, 3rd ed. (1845), p. 268.

13. Karolevitz, *Newspapering*, p. 148.

The Foster Press

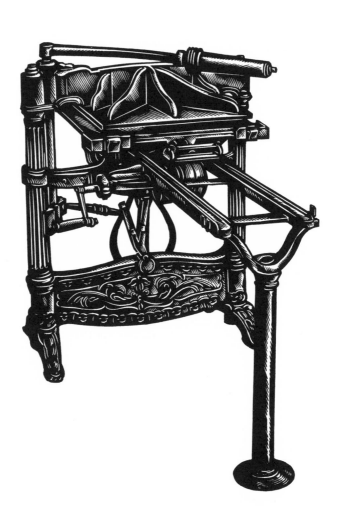

) 12 (

The Foster Press
1850–1868

THE FOSTER PRESS seems to be the only surviving example of an odd variety of press. The line seems to have started about 1836 when James Maxwell, a New York machinist, invented the Eagle press. An advertisement in the 1836 edition of Van Winkle's *The Printers' Guide*[1] shows a press that looked similar to the Ruthven. The Eagle was a half-height press with what appears to be a simple Wells equal-length lever arrangement placed under the bed. When the bed was rolled in under the platen, a pull of the bar pushed the bed upward (instead of the platen downward), to make the impression. The Van Winkle advertisement states that Maxwell was the ''inventor and patentee,'' but his name is not on any patent that can be found.

Ralph Green has written, ''a man named Jones, living in Cincinnati, revived the press in 1851, and advertised it as an entirely new idea in press construction.''[2] Green gives no further information about Jones or his source for this information except to mention that it was made by Guilford & Jones in Cincinnati.[3] Then we reach somewhat firmer ground. The year after the 1851 revival, Charles Foster of Cincinnati received a patent, dated October 5, 1852, for a press very similar to the original Eagle. Green says that Foster ''had been working for the Cincinnati Type Foundry as a machinist in the hand press department.''[4] We do know that he was the principal of C. Foster & Co., ''Dealers in

Printers' supplies; Printing-press manufacturers (Foster's Steam Press)" from 1845 to 1849; the company was succeeded by C. Foster & Bro. from 1850 to 1852 and by James D. Foster & Co. in 1856.[5]

The 1852 patent shows a half-height press with a variation of the Washington-type toggle under the bed. The innovation claimed in the patent dealt mainly with the rails.

As the bed of the press was drawn under the platen, it moved on inclined rails which moved it upward and so reduced the distance (and presumably the effort) needed to make an impression. The other claim was for rails that were pivoted at the end away from the press, so that when the bar was pulled the rails, bed and form were raised together against the stationary platen for the impression. Releasing the bar caused the rails to pivot down again under the weight of the bed, thus obviating the need for springs.

"Foster's Improved Hand Printing Press" was illustrated and described on the first page of an 1853 issue of *Scientific American*.[6] Although the press shown is very similar to the press illustrated here, which is at the Henry Ford Museum in Dearborn, Michigan, there are some differences. The press shown in *Scientific American* is more ornate, with a filigree casting at the top and brass finials surmounting the fluted columns at the sides. In contrast, the Foster press at the Ford Museum lacks the filigree and finials, but it is essentially the same in shape and arrangement.

By the time of the *Scientific American* article, Foster had moved from Cincinnati to Philadelphia, where the press was produced at No. 1, Lodge Alley.[7] It was exhibited at the Crystal Palace exhibition in New York in 1853, and it seems to have had some success in the years following. Six years later, it was one of four presses manufactured by Foster that were shown in a full page advertisement in the type specimen book of the New York Type Foundry (Charles White & Co.) [8] and, in 1862, in the specimen book of the Chicago Type Foundry[9] (generally known today as the Marder, Luse foundry). The two foundries had the same ownership, and the advertisements are identical. The illustration shows slight changes from the *Scientific American* picture. Although the moving rails are still incorporated, they are not mentioned in the advertising copy. Emphasized were the increase of power, simplicity of construction, counterbalanced tympan, and the under-the-bed mechanism— none of which were innovative features.

Foster's other presses, shown in the same advertisement, are two platen jobbers similar to Gordon's "Alligator Press" and a crank-operated patent card press. In addition to these presses, we have reports of a more traditional hand press made by Foster. One is located in New Brunswick, New Jersey and bears a great resemblance to the press at Shelburne Museum (*q.v.*) in the shape of its frame.[10] Another traditional Washington-style Foster hand press has been located in the office of the *Opp News*, Opp, Alabama. It was built by "C. Foster & Co., Cincinnati" according to the inscription cast on the press frame.[11]

"Foster's Patent Hand Press" was sold in seven sizes, as follows: (Sizes given are of the platen.)

Foolscap	14 x 18	$95
Medium	19 x 25	$150
Super Royal	22–1/2 x 29	$190
Imperial	23 x 32	$195
Double Medium	24–1/2 x 38–1/2	$200
Imperial, No. 5	26 x 40	$210
Double Super Royal	28 x 42–1/2	$220[12]

The press illustrated here is the Imperial size. As mentioned above, it is in the collection of the Henry Ford Museum. It stands about 35 inches high to the top of the rails. It was acquired by Henry Ford for his museum in 1929 from W.A. Starr of the Goshen Printery, Goshen, Indiana. Mr. Starr had bought it sight unseen for use as a proof press, thinking that he was buying a standard Washington press.[13] He was startled to find that he had bought what he referred to as "this monstrosity."[14]

The only other press like it that we know of is in the collection of the Findlay Historical Society in Findlay, Ohio. The Findlay press was originally used to print the Findlay *Weekly Jeffersonian* during the Civil War. It was on this press[15] that the famous letters of "Petroleum V. Nasby"[16] were printed, which so amused President Lincoln and the American public.

Foster's patent was bought out by the R. Hoe and Co. in 1857,[17] a fact which would indicate their expectation of success in the marketplace. Hoe used the patent to design their "Improved Washington Press" which they hoped would supplant the standard press. That did not happen, and Hoe discontinued the Improved Washington in 1868.[18]

NOTES FOR CHAPTER TWELVE

1. Cornelius Van Winkle, *The Printers' Guide*, 3rd ed. (New York, 1836), p. [242]; Green, *Iron Hand Press*, pp. 25-26.

2. Green, *Iron Hand Press*, p. 26.

3. *Ibid.*, p. 29.

4. *Ibid.*, p. 26.

5. Sutton, *Western Book Trade*, pp. 322-323.

6. "Foster's Improved Hand Printing Press," *Scientific American* Vol. 9, No. 16 (New York, Dec. 31, 1853), p. 121.

7. *Idem.*

8. *Specimen Book of Printing Types*, Chas. T. White & Co. (New York, 1858), p. [174].

9. *Specimen Book of Printing Types*, Chicago Type Foundry (Chicago, 1862), p. [253].

10. Letter from Elizabeth Harris, Smithsonian Institution, 4 December 1986.

11. Letters from Walter Clement, 10 December 1986 and 15 January, 1987.

12. *Specimen Book of Printing Types*, Charles T. White & Co. (New York, 1858), p. [174].

13. Letter from W.S. Starr to Henry Ford, 15 June 1929, at the Henry Ford Museum, Dearborn Michigan.

14. Letter from W.S. Starr to Henry Ford, 23 December 1929, at the Henry Ford Museum.

15. R.L. Heminger, *Hancock County and the Civil War* (Findlay, Ohio, 1961?), p. [29]. See also R.L. Heminger, *Across the Years in Findlay and Hancock County* (Findlay, Ohio, 1965), pp. 41-42.

16. pseud. for David Ross Locke (1833-1888).

17. Green. p. 26.

18. *Idem.*

The Ruggles Press

The Ruggles Press
1859–?

AMID THE BOOKS on printing and graphic arts deep in the stacks of Harvard's rare book collection, The Houghton Library, stands an iron hand press with a brass plate that reads:

PATENTED
May 19, 1859
and
Manufactured
by
S. P. RUGGLES
Boston
Mass.

It is the only known surviving hand press made by the Boston inventor and press maker, Stephen P. Ruggles.

The date of the patent is quite late as hand presses go, but it indicates that there was still a market for these presses at a time when power presses were well-established in the larger offices and platen jobbers were coming on the scene.

Stephen P. Ruggles (1808-1880) had been making presses for many years before this. While he was still an apprentice printer at the *Vermont Republican and American Yeoman* he made an early attempt to ink the type on a hand press with a roller. When hardly "out of his time" (that is, just after his apprenticeship ended) in 1826 he moved to Boston and invented and built a large cylinder power press on which was printed in Boston *The Ladies' Magazine*.[1]

In 1830-31, at George Minor's machine shop in New York, Ruggles invented and built the first card press, which proved to be a mechanical and financial success for Ruggles. From 1833 to 1838 he became involved in producing inventions for the education of the blind, including presses powerful enough to emboss letters into paper. His invention substantially lowered costs, and made possible the production of an American Bible for the blind.[2] After 1838 he returned to devising new forms of printing presses. His embossing press was adapted commercially in 1839 as the Ruggles Printing Engine and was very successful.[3]

Ruggles invented and sold several novel presses, including the Printing Engine, the Rotary Card and Job Press, and the Combination Job Press. These platen jobbers, operated by foot treadles, were the most successful of their kind until the mid-1850s, when George Gordon's Franklin Press appeared.

Ruggles' 1859 hand press is interesting because it returns to the screw principle as a means of applying pressure. At the top of the press, contained inside a cylinder, was a double-threaded screw surrounded by a coil spring. A coarse screw was used to effect the raising and lowering of the platen. After the coarse screw had run its course, the action of the finely-threaded screw began and completed the actual impression. The patent claim describes the action thus:

> The object of the two screws, viz. the coarse and fine one, both actuated by the lever F, is that a slight motion of the lever F, say a quarter turn, will raise the platen sufficiently high from the bed, to allow the form, frisket, sheet, blankets, and the appliances generally used in connection therewith, to freely run under

without slurring the sheets, while the fine screw comes into action just at the time when the greater power is required to take the impression.[4]

Ruggles added that the transition from the action of one screw to the other was undetectable, and the device could be adjusted to begin and end the action of each screw as desired. Ruggles also claimed another improvement. The bed of his press ran on inclined rails, like the Foster press. As the bed was moved in under the platen, it simultaneously traveled upward on the rails. The effect of this was to utilize the horizontal motion of the press to assist the vertical approach of the platen to the bed.

A requirement for a U.S. patent from 1836 to about 1870 was the submission of a working model. A beautifully-made model of the Ruggles press is on view in the graphic arts exhibition area of the Smithsonian Institution.

In 1854 Ruggles sold all his printing press patents and the good will of his business to the S.P. Ruggles Power Press Manufacturing Co. and retired with a fortune of over $500,000, just as the platen presses of George P. Gordon were sweeping the field.[5] Gordon acknowledged that the basic principles employed in his presses were invented by Ruggles.[6]

The Houghton Library Ruggles hand press is quite small, with a platen measuring only 13″ x 19–1/2″. The frame is a horizontal oval, and the projecting rails are unsupported. There are no records at the Houghton Library to indicate how it got there, although there is a surmise that Philip Hofer may have purchased it.[7] Apart from the patent model at the Smithsonian, no other Ruggles hand press is known.

NOTES FOR CHAPTER THIRTEEN

1. "The S.P. Ruggles' Power Press Manufacturing Company," *Ballou's Pictorial Drawing-Room Companion* (Boston, 30 June 1855), p. 405; Rollo G. Silver, "The Autobiography of Stephen P. Ruggles," *Printing History* 1, Vol. 1, No. 1 (New York, 1979), pp. 10-11.

2. Comparato, *Chronicles*, pp. 174-175; Ralph Green, *A History of the Platen Jobber* (Chicago., 1953), pp. 4-6.

3. Moran, *Printing Presses*, pp. 146-147; Green, *Platen Jobber*, p. 9.

4. U.S. Patent Office, "Specification of Letters Patent No. 23,951, dated May 10, 1859."

5. Silver, *Autobiography of Stephen P. Ruggles*, pp. 7, 17.

6. Letter dated 11 October 1874 from George P. Gordon to Ruggles, reprinted in MacKellar, Smiths & Jordan's *Typographic Advertiser*, Vol. XXIV, Nos. 95/96 (Springtime, 1879), p. 689. The letter is mentioned, without details, in Green, *Platen Jobber*, p. 12.

7. Letter from Eleanor M. Garvey, Houghton Library, 7 October 1987.

Other Presses

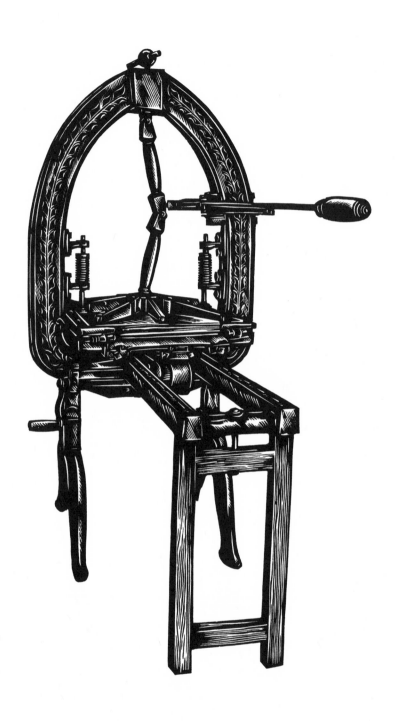

❧ 14 ❧

Other Presses
The Press at the Shelburne Museum
ca. 1825

THIS PRESS is something of a mystery. Originally it was believed
to be the press patented in 1827 by Samuel Couillard, Jr. of Boston. There are two objections to this: the press looks dissimilar to
the Couillard, as shown in the patent drawing and in a wood engraving published in the 1828 *An Abridgment of Johnson's "Typographia."*[1] Second, the Shelburne press lacks the distinguishing
feature claimed in Couillard's patent: the use of a friction wheel
on the frame of the press, which rotates against a cam placed on
the bar of the press. At present it seems likely that there is no surviving Couillard press.

The press at the Shelburne Museum in Vermont is the only one
of its kind known, and it is illustrated here (opposite). The frame
is a gothic arch very similar to the Union press (*q.v.*) but it is decorated with a leaf border instead of stars. (The shape of the frame
provides one tenuous link with Couillard. He received a patent
for a hand press with a gothic arch frame on October 5, 1827. The
press is otherwise quite different from the Shelburne press.)

The press came from L.S. Bolster, a printer in Chester, Vermont
who acquired it as part of the Chester Printing Company in 1925.
The Chester Printing Company had been in business since 1888
and it seems likely that this press was part of its equipment. Mr.
Bolster wrote to Ralph Green about the press with a request for
information, but Mr. Green replied that he had never encountered

the press before. He judged it to date from about 1825, based on the toggle mechanism.[2] He later listed it in his *The Iron Hand Press in America* but did not discuss it.[3]

The press was purchased from the Bolster estate in 1959 for $750 by printer Frank Teagle of Woodstock, Vermont. It was repaired by George Sharon of the same city and was later given by Mr. Teagle to Brad Brownell, a printer and curator of the Ben Lane Print Shop at Shelburne, with the stipulation that it later be given to some museum. When Brad Brownell died in 1987 the press remained on view at the Shelburne Museum.

There are no markings on the press except for the leaf border ornamentation. It is possible that a brass plate with information was once affixed to the vertical rectangle at the top of the frame. The platen measures 27″ x 21″, and the press is 6′–0″ tall.

The Couillard Press
1827

The Couillard press seems not to have survived the ravages of time. What is known about it is revealed in the patent papers and terse references and a picture in an 1828 printers' manual.

The press was invented by Samuel Couillard, Jr. of Boston, who patented it July 14, 1827. The mechanism of the press was an equal-length toggle like the one patented years before by John Wells. A pull of the bar straightened the two parts of the toggle into a vertical line, thus forcing the platen downward for the impression. The patent claim referred mainly to a friction-wheel placed at the side of the press, which moved against a cam on the bar of the press near the pivot-point, helping to steady the movement of the bar and the platen. Another claim related to the curvature of the bar, which Couillard said provided for a "dwell" during the impression. The patent did not go beyond these minor improvements.

The patent drawing shows an oval frame with a flattened top. The only other illustration of the press shows essentially the same press, with an ornamental finial added at the top. It appears in *An Abridgment of Johnson's "Typographia,"* published in Boston in 1828. Accompanying the wood engraving is the brief statement, "Invented by Samuel Couillard, Jun., and Manufactured by the Proprietor, P. Dow, Boston. Mr. John G. Rogers, Agent."[4] The same wood engraved illustration, unlabeled, appears a few years later as cut number 489 offered for sale in the 1832 specimen book of the Boston Type and Stereotype Foundry.[5] John Gorham Rogers, mentioned as agent for the press, held the same position for the Boston Type and Stereotype Foundry, which helps to explain the presence of the cut in the foundry's specimen book.

Couillard does not seem to have been associated with the manufacture of the press he invented. He has later patents to his credit relating to the cleaning and dyeing of wool (1833) and the combing of flax and hemp (1836). The Couillard press was manufactured by Phineas Dow (1780-1869?), an expert Boston machinist who built fire engines and who became something of a specialist in making and repairing printing presses. Dow made the first model and the working presses for inventor Daniel Treadwell. He employed in his shop the brothers Seth and Isaac Adams, as well as Otis Tufts, all of whom later invented or manufactured iron hand presses.[6] Thus Dow was the common denominator of most of the early press manufacturing in Boston. His employee Isaac Adams was the inventor of the Union press (and had been the apprentice of Erastus Bartholomew, who manufactured it.)[7]

The Adams Press
1831–1859

There is no mystery about the hand presses made by Isaac and Seth Adams of Boston. Although Isaac Adams was the inventor of the Union press (*q.v.*) and the famous Adams Power Press, he also made and sold hand presses that were copies of other inventors' work. These presses are scattered about New England in old newspaper offices and in collections.

The Adams presses were always in the acorn frame made popular by Peter Smith's press. Many made use of the equal-length toggle mechanism as well. A good example of one of these presses is on view at the Updike Printing Collection at the Providence Public Library. It is thought to have been used to print the Pawtucket *Chronicle* in the early part of the nineteenth century.[8] A similar Adams press is on view at the Vineyard *Gazette* on Martha's Vineyard, Massachussetts. It was used to print that newspaper from its first issue in 1846, and in the era of the Linotype it was used as a proof press.[9] Yet another of these Adams presses is at the Bristol *Phoenix*, Bristol, Rhode Island.

Adams also copied the Stansbury torsion toggle mechanism, usually for smaller proof presses. The familiar Smith acorn frame was used for these presses as well. Examples of these smaller presses can be seen at the Updike Collection; and another at E.L. Freeman Company, both in Providence.

Nameplates on the presses vary, and include I. Adams & Co.; I. & S. Adams, and Seth Adams & Co., all of Boston. For more about Isaac Adams, see the chapters on the Union and Tufts presses.

The following early presses have been mentioned by Green.[10] A few of the presses in this group have survived, as noted.

AMERICAN PRESS. Made by Adam Ramage, about 1844. A variation of the Washington frame and toggle. Pictured in Adams' *Typographia*, 1844, where it is stated it was invented by Shelden Graves of Philadelphia.[11] It is possible that some examples survive, but they are not known at present.

AUSTIN PRESS. Patented by Frederick J. Austin of New York, October 8, 1836. One of these presses, formerly owned by Morgan Press, Hastings-on-Hudson, New York, is now in Idaho. Another is presently in Massachusetts.

COSFELDT PRESS. Mentioned by Adams as having been a Washington-style press made by F. J. Cosfeldt of Philadelphia.[12] No surviving examples are known.

FRANKLIN PRESS. Green describes this press as having a cast iron harp-shaped frame, and says it was made by Dickinson & Williamson of Cincinnati, *ca.* 1845.[13] The source of Green's information is not known. One of these presses is located at the Old Fort Museum, Mackinac Island, Michigan. A photograph of this press can be found in one of Ralph Green's scrapbooks at the Alderman Library, College of William and Mary, Williamsburg.

EAGLE PRESS. This half-height press was similar to the Foster except that it used Wells' equal length levers under the bed. It was advertised in the 1836 edition of Van Winkle's *Printers' Guide*. The advertisement pictured the press and stated that it was manufactured by James Maxwell of New York in four sizes. No Eagle press has survived.

NOTES FOR CHAPTER FOURTEEN

1. [Thomas F. Adams], *An Abridgment of Johnson's "Typographia"* (Boston, 1828), p. 308.

2. Information about the provenance of the press in a letter from Frank Teagle, 20 March 1986.

3. Green, *Iron Hand Press*, p. 28.

4. [Adams], *Abridgment*, pp. 308-309.

5. *Specimen of Printing Types from the Boston Type & Stereotype Foundry* (Boston, 1832), p. [338]. Facsimile edition edited by Stephen O. Saxe (New York: Dover, 1989), p. 156.

6. Comparato, *Chronicles*, pp. 15-19.

7. Bartholomew, *Record*, p. 174; "Isaac Adams," *Ballou's*, 24 Feb. 1855, p. 124.

8. "Adams Acorn Press," (Providence, 1984?) (information sheet distributed by Daniel Berkeley Updike Collection on Printing, Providence Public Library.

9. Vineyard *Gazette*, 27 April 1984, p. 3B.

10. Green, *Iron Hand Press*, p. 28.

11. Adams, *Typographia*, 2nd ed., (Philadelphia, 1844), p. 268.

12. *Ibid*.

13. Ralph Green, unpublished working list dated 4 February 1942.

Afterword

MUCH MORE remains to be written about the iron hand press in North America. One area that needs exploration is the interconnected group of machinists, press inventors and manufacturers who worked in Boston in the late 1820s and 1830s. Their presses, including the Union, Couillard, Adams and Tufts, are much less documented than the others. The press at the Shelburne Museum seems to be closely related to the Union press, but its identity is still unknown.

I believe, however, that in this book I have collected much of what is known of iron hand presses in this country. Where space did not permit a complete description, sources have been cited for greater detail. While presses like the Washington, Columbian and Albion are well known, several more obscure presses have been described. These include not only the Boston presses mentioned above, but also some others hardly mentioned by Ralph Green and other writers: the Ruthven, Stansbury, Philadelphia, and Foster presses. I am not aware of any contemporary description of the Ruggles or Union presses, the missing Couillard press, or the unidentified press at Shelburne.

It is my hope that this book will elicit more information about these presses. I will be grateful for such information, as well as for any corrections that may be called for.

Appendix A

PRESS	PATENTEE	RESIDENCE	DATE
Ruthven	John Ruthven	Edinburgh	*Nov. 1, 1813
Columbian	George Clymer	London	*Nov. 1, 1817
	Adam Ramage	Philadelphia	May 28, 1818
Wells	John I. Wells	Hartford	Feb. 8, 1819
Stansbury	Abraham O. Stansbury	New York	April 7, 1821
Washington	Samuel Rust	New York	May 13, 1821
	David Barclay	London	*July 26, 1821
Smith	Peter Smith	New York	Dec. 29, 1821
	Peter Smith	New York	April 6, 1822
	Adam Ramage	Philadelphia	May 19, 1823
	David Phelps	Boston	Sept. 15, 1826
Couillard	Samuel Couillard, Jr.	Boston	July 14, 1827
	Samuel Couillard, Jr.	Boston	Oct. 5, 1827
Washington	Samuel Rust	New York	April 17, 1829
	John I. Wells	Hartford	June 29, 1829
Tufts	Otis Tufts	Boston	July 30, 1831
Tufts	Otis Tufts	Boston	Nov. 7, 1831
	Otis Tufts	Boston	Aug. 22, 1834
	Adam Ramage	Philadelphia	Nov. 19, 1834
Eagle	James Maxwell	New York	? ca. 1836
Foster	Charles Foster	Cincinnati	Oct. 5, 1852
Ruggles	Stephen P. Ruggles	Boston	May 10, 1859

Sources: *Abridgments of Specifications Relating to Printing* (London, 1859); *A List of Patents Granted by the United States from April 10, 1790 to December 31, 1836* (Washington, 1872); *Early Unnumbered United States Patents 1790 -1836* (Woodbridge, Conn. 1980); *Subject Matter Index of Patents and Inventions, 1790 -1873* (Washington, 1873). British patents are marked with an asterisk, and are not intended to be complete.

Appendix B

Before 1875	Area sq. in.	After 1875	Area sq. in.
		8 x 12 & 9 x 12	96-108
Foolscap 18x14	252	14x18	252
Foolscap 19½x14½	283		
Crown 21x16	336	16x21	336
Crown 22x16	352		
		17x21	357
Demy 24x18	432		
Medium 25x19	475	5 column folio 19x25	475
Medium 25½x19½	497	5 column folio 20x25	500
Royal 26x20	520		
Royal 26x20½	533		
Super Royal 28x21	588		
Super Royal 29½x23	678		
Imperial No.1 30x21	630	6 column folio 21x30	630
		6 column folio 23x31	713
Imperial No.2 32½x22	715	6 column folio 22x32½	715
Imperial No.3 32x23	736		
Imperial No.3 35x23	805	7 column folio 23x35	805
Imperial No.4 37x24	888	7 column folio 24x37	888
Imperial No.4 38x24	912		
		8 column folio 25x38	950
Imperial No.5 39x25	975	8 column folio 25x39	975
Imperial No.5 39x26	1014		
Imperial No.6 41½ x26	1079		
Imperial No.6 41¾ x28¼	1179		
Imperial No.7 43x27	1161	9 column folio 27x43	1161

	Area sq. in.		Area sq. in.
		9 column folio 28½ x 43	1225
		6 column quarto 29 x 43	1247
Imperial No.7 47 x 31	1457		
Mammoth 43¼ x 34¼	1481		
Mammoth 43½ x 34½	1501		
Imperial No.8 50 x 33	1650		
Imperial No.9 53 x 36	1908		
Imperial No.9 56 x 37	2072		

Before 1875 the sizes of hand presses were given in terms of the size of the sheet of paper they could print. The platen size was often also listed in advertisements, and sometimes the size of the bed. After 1850 hand presses were only used for newspapers of limited circulation, such as country weeklies. It was not until 1875 that the size of a hand press was expressed in terms of the size of the newspaper the press could print.

Sizes of early treadle platen presses were based on a fractional part of the Medium hand press. Presses from 7 x 11 to 9 x 13 were classed as "⅛ medium"; from 10 x 15 to 12 x 18 as "quarto medium"; and from 13 x 19 to 17 x 22 as "half medium."

Appendix C

		Area sq. in.
Royal Octavo	10x7	70
Foolscap Folio	15x9¾	146
Post Folio	16x11	176
Demy Folio	18x12	216
Foolscap Broadside	19x14½	275
Crown	21x16	336
Demy	24x18	432
Royal	26x20½	533
Super Royal	29x21	609
Double Crown	34x22½	765
Double Demy	36x23	828
Double Royal	40x25	1000
Extra Size Double Royal	42x27	1134

Sources: The chart of American hand press sizes and text accompanying it are based on unpublished material prepared by Ralph Green, 1/15/46, in Notebooks of Glover Snow, The Kelsey Company, Meriden, Connecticut. The chart of British hand press sizes is based on a list in *Revised Illustrated Price List of New Machinery and Materials* issued by Frederick Ullmer, Ltd., May 1902, pp. 41-43.

Selected Bibliography

[ADAMS, THOMAS F.] *An Abridgment of Johnson's "Typographia."* Boston 1828.

ADAMS, THOMAS F. *Typographia.* Philadelphia, 1837. 2nd edition, Philadelphia, 1844; 3rd edition, Philadelphia, 1845.

BIGMORE, E.C. AND WYMAN, C.W.H. *A Bibliography of Printing.* London, 1880-1886.

BULKELEY, HOUGHTON. "John I. Wells, Cabinetmaker-Inventor," *Connecticut Historical Society Bulletin,* Vol. 26 No. 3, July, 1961. Hartford, 1961.

COMPARATO, FRANK E. *Chronicles of Genius and Folly.* Culver City, California, 1979.

GREEN, RALPH. *The Iron Hand Press in America.* Rowayton, Connecticut, 1948. Reprinted Cincinnati, 1981.

HAMILTON, MILTON W. *Adam Ramage and His Presses.* Portland, Maine, 1942.

HANSARD, T.C. *Typographia.* London, 1825.

HARRIS, ELIZABETH. "Press-builders in Philadelphia, 1776-1850," *Printing History* 22, Vol. XI, No. 2. New York, 1989.

HART, HORACE. *Charles, Earl Stanhope and the Oxford University Press.* Reprinted from Collecteana III, 1896, of the Oxford Historical Society, with notes by James Mosley. London, 1966.

JOHNSON, JOHN. *Typographia.* London, 1824.

KAINEN, JACOB. *George Clymer and the Columbian Press.* New York, 1950.

LOGAN, HERSCHEL. *The American Hand Press.* Whittier, California, 1980.

MORAN, JAMES. *Printing Presses.* Berkeley and Los Angeles, 1973.

MORAN, JAMES. "The Columbian Press," *Journal of the Printing Historical Society,* No. 5. London, 1969.

OLDHAM, ROBERT. "Abraham O. Stansbury and the Torsion Toggle Hand Printing Press," *The Chronicle of the Early American Industries Association* Vol. 36, No. 4, December, 1983. Albany, 1983.

[PASKO, W.W.] *American Dictionary of Printing and Bookmaking.* New York, 1894.

RINGWALT, J. LUTHER. *American Encyclopaedia of Printing.* Philadelphia, 1871.

SAVAGE, WILLIAM. *Dictionary of the Art of Printing.* London, 1841.

SILVER, ROLLO G. "The Autobiography of Stephen P. Ruggles." *Printing History* 1, Vol. 1, No.1. New York, 1979.

SILVER, ROLLO G. *The American Printer, 1787-1825.* Charlottesville, Virginia, 1967.

STONE, REYNOLDS. "The Albion Press." *Journal of the Printing Historical Society,* No. 2. London, 1966.

STOWER, CALEB. *The Printer's Grammar.* London, 1808.

VAN WINKLE, CORNELIUS. *The Printers' Guide.* New York, 1818. 2nd edition, New York, 1827; 3rd edition, New York, 1836.

WILKES, WALTER. *Die Entwicklung der Eisenen Buchdruckerpresse.* Pinneberg, West Germany, 1983.

Index

INDEX

American Iron Hand Presses was designed by Neil Shaver at the Yellow Barn Press, Council Bluffs, Iowa. Reproduction proofs of the original letterpress edition were supplied to Oak Knoll Books for this trade edition which has been published in both hard and soft cover. All copies have been printed on archival quality paper. The text is set in 14 point Monotype Bulmer.